MODULAR LOW-POWER, HIGH-SPEED CMOS ANALOG-TO-DIGITAL CONVERTER FOR EMBEDDED SYSTEMS

THE KLUWER INTERNATIONAL SERIES IN ENGINEERING AND COMPUTER SCIENCE

ANALOG CIRCUITS AND SIGNAL PROCESSING

Consulting Editor*: Mohammed Ismail. *Ohio State University

Related Titles:

DESIGN CRITERIA FOR LOW DISTORTION IN FEEDBACK OPAMP CIRCUITE
Hernes & Saether
ISBN: 1-4020-7356-9

CIRCUIT TECHNIQUES FOR LOW-VOLTAGE AND HIGH-SPEED A/D CONVERTERS
Walteri
ISBN: 1-4020-7244-9

DESIGN OF HIGH-PERFORMANCE CMOS VOLTAGE CONTROLLED OSCILLATORS
Dai and Harjani
ISBN: 1-4020-7238-4

CMOS CIRCUIT DESIGN FOR RF SENSORS
Gudnason and Bruun
ISBN: 1-4020-7127-2

ARCHITECTURES FOR RF FREQUENCY SYNTHESIZERS
Vaucher
ISBN: 1-4020-7120-5

THE PIEZOJUNCTION EFFECT IN SILICON INTEGRATED CIRCUITS AND SENSORS
Fruett and Meijer
ISBN: 1-4020-7053-5

CMOS CURRENT AMPLIFIERS; SPEED VERSUS NONLINEARITY
Koli and Halonen
ISBN: 1-4020-7045-4

MULTI-STANDARD CMOS WIRELESS RECEIVERS
Li and Ismail
ISBN: 1-4020-7032-2

A DESIGN AND SYNTHESIS ENVIRONMENT FOR ANALOG INTEGRATED CIRCUITS
Van der Plas, Gielen and Sansen
ISBN: 0-7923-7697-8

RF CMOS POWER AMPLIFIERS: THEORY, DESIGN AND IMPLEMENTATION
Hella and Ismail
ISBN: 0-7923-7628-5

DATA CONVERTERS FOR WIRELESS STANDARDS
C. Shi and M. Ismail
ISBN: 0-7923-7623-4

DIRECT CONVERSION RECEIVERS IN WIDE-BAND SYSTEMS
A. Parssinen
ISBN: 0-7923-7607-2

AUTOMATIC CALIBRATION OF MODULATED FREQUENCY SYNTHESIZERS
D. McMahill
ISBN: 0-7923-7589-0

MODEL ENGINEERING IN MIXED-SIGNAL CIRCUIT DESIGN
S. Huss
ISBN: 0-7923-7598-X

ANALOG DESIGN FOR CMOS VLSI SYSTEMS
F. Maloberti
ISBN: 0-7923-7550-5

CONTINUOUS-TIME SIGMA-DELTA MODULATION FOR A/D CONVERSION IN RADIO RECEIVERS
L. Breems, J.H. Huijsing
ISBN: 0-7923-7492-4

DIRECT DIGITAL SYNTHESIZERS: THEORY, DESIGN AND APPLICATIONS
J. Vankka, K. Halonen
ISBN: 0-7923 7366-9

SYSTEMATIC DESIGN FOR OPTIMISATION OF PIPELINED ADCs
J. Goes, J.C. Vital, J. Franca
ISBN: 0-7923-7291-3

MODULAR LOW-POWER, HIGH-SPEED CMOS ANALOG-TO-DIGITAL CONVERTER OF EMBEDDED SYSTEMS

by

Dr.-Ing. Keh-La Lin

National Cheng Kung University (NCKU), Tainan, Taiwan

Armin Kemna

Fraunhofer Institute (FhG-IMS), Duisburg, Germany

and

Prof. Bedrich J. Hosticka

Fraunhofer Institute (FhG-IMS), Duisburg, Germany and *Gerhard-Mercator-University, Duisburg*

SPRINGER SCIENCE+BUSINESS MEDIA, LLC

A C.I.P. Catalogue record for this book is available from the Library of Congress.

ISBN 978-1-4419-5340-7 ISBN 978-0-306-48726-2 (eBook)
DOI 10.1007/978-0-306-48726-2

Printed on acid-free paper

Originally published by Kluwer Academic Publishers, Boston in 2003
Softcover reprint of the hardcover 1st edition 2003

Dedicated to our families.

FORWORD

This book results from of a large European project started in 1997, whose goal is to promote the further development and the faster and wider industrial use of advanced design methods for reducing the power consumption of electronic systems.

Low power design became crucial with the wide spread of portable information and communication terminals, where a small battery has to last for a long period. High performance electronics, in addition, suffers from a permanent increase of the dissipated power per square millimeter of silicon, due to the increasing clock-rates, which causes cooling and reliability problems or otherwise limits the performance.

The European Union's Information Technologies Programme 'Esprit' did therefore launch a 'Pilot action for Low Power Design', which eventually grew to 19 R&D projects and one coordination project, with an overall budget of 14 million EURO. It is meanwhile known as European Low Power Initiative for Electronic System Design (ESD-LPD) and will be completed in the year 2002. It involves to develop or demonstrate new design methods for power reduction, while the coordination project takes care that the methods, experiences and results are properly documented and publicised.

The initiative addresses low power design at various levels. This includes system and algorithmic level, instruction set processor level, custom processor level, RT-level, gate level, circuit level and layout level. It covers data dominated and control dominated as well as asynchronous architectures. 10 projects deal mainly with digital, 7 with analog and mixed-signal, and 2 with software related aspects. The principal application areas are communication, medical equipment and e-commerce devices.

The following list describes the objectives of the 20 projects. It is sorted by decreasing funding budget.

CRAFT	CMOS Radio Frequency Circuit Design for Wireless Application

- Advanced CMOS RF circuit design including blocks such as LNA, down converter mixers & phase shifters, oscillator and frequency synthesiser, integrated filters delta sigma conversion, power amplifier
- Development of novel models for active and passive devices as well as fine-tuning and validation based on first silicon fabricates
- Analysis and specification of sophisticated architectures to meet in particular low power single chip implementation

PAPRICA Power and Part Count Reduction Innovative Communication Architecture

- Feasibility assessment of DQIF, through physical design and characterisation of the core blocks
- Low-power RF design techniques in standard CMOS digital process
- RF design tools and framework; PAPRICA Design Kit.
- Demonstration of a practical implementation of a specific application

MELOPAS Methodology for Low Power ASIC Design

- To develop a methodology to evaluate the power consumption of a complex ASIC early on in the design flow
- To develop a hardware/software co-simulation tool
- To quickly achieve a drastic reduction on the power consumption of electronic equipment

TARDIS Technical Coordination and Dissemination

- To organise the communication between design experiments and to exploit their potential synergy
- To guide the capturing of methods and experiences gained in the design experiments
- To organise and promote the wider dissemination and use of the gathered design know-how and experience

LUCS Low-Power Ultrasound Chip Set

- Design methodology on low power ADC, memory and circuit design
- Prototype demonstration of a handheld medical ultrasound scanner

ALPINS Analog Low Power Design for Communications Systems

- Low-voltage voice band smoothing filters and analog-to-digital and digital-to-analog converters for an analog front-end circuit of a DECT system

- High linear transconductor-capacitor (gm-C) filter for GSM Analog Interface Circuit operating at supply voltages as low as 2.5V
- Formal verification tools, which will be implemented in the industrial partners design environment. These tools support the complete design process from system level down to transistor level

SALOMON System-level analog-digital trade-off analysis for low power

- A general top-down design flow for mixed-signal telecom ASICs
- High-level models of analog and digital blocks and power estimators for these blocks
- A prototype implementation of the design flow with particular software tools to demonstrate the general design flow

DESCALE Design Experiment on a Smart Card Application for Low Energy

- The application of highly innovative handshake technology
- Aiming at some 3 to 5 times less power and some 10 times smaller peak currents compared to synchronously operated solutions

SUPREGE A low power SUPerREGEnerative transceiver for wireless data transmission at short distances

- Design trade-offs and optimisation of the micro power receiver / transmitter as a function of various parameters (power consumption, area, bandwidth, sensitivity, etc)
- Modulation / demodulation and interface with data transmission systems
- Realisation of the integrated micro power receiver / transmitter based on the super-regeneration principle

PREST Power REduction for System Technologies

- Survey of contemporary Low Power Design techniques and commercial power analysis software tools
- Investigation of architectural and algorithmic design techniques with a power consumption comparison
- Investigation of Asynchronous design techniques and Arithmetic styles
- Set-up and assessment of a low power design flow
- Fabrication and characterisation of a Viterbi demonstrator to assess the most promising power reduction techniques

DABLP Low Power Exploration for Mapping DAB Applications to Multi-Processors

- A DAB channel decoder architecture with reduced power consumption
- Refined and extended ATOMIUM methodology and supporting tools

COSAFE Low Power Hardware-Software Co-Design for Safety-Critical Applications

- The development of strategies for power efficient assignment of safety critical mechanisms to hardware or software
- The design and implementation of a low-power, safety-critical ASIP, which realises the control unit of a portable infusion, pump system

AMIED Asynchronous Low-Power Methodology and Implementation of an Encryption/Decryption System

- Implementation of the IDEA encryption/decryption method with drastically reduced power consumption
- Advanced low power design flow with emphasis on algorithm and architecture optimisations
- Industrial demonstration of the asynchronous design methodology based on commercial tools

LPGD A Low-Power Design Methodology/Flow and its Application to the Implementation of a DCS1800-GSM/DECT Modulator/Demodulator

- To complete the development of a top-down, low power design methodology/flow for DSP applications
- To demonstrate the methods at the example of an integrated GFSK/GMSK Modulator-Demodulator (MODEM) for DCS1800-GSM/DECT applications

SOFLOPO Low Power Software Development for Embedded Applications

- Develop techniques and guidelines for mapping a specific algorithm code onto appropriate instruction subsets
- Integrate these techniques into software for the power-conscious ARM-RISC and DSP code optimisation

I-MODE Low Power RF to Base band Interface for Multi-Mode Portable Phone

- To raise the level of integration in a DECT/DCS1800 transceiver, by implementing the necessary analog base band low-pass filters and data converters in CMOS technology using low power techniques

COOL-LOGOS Power Reduction through the Use of Local don't Care Conditions and Global Gate Resizing Techniques: An Experimental Evaluation.

- To apply the developed low power design techniques to the existing 24-bit DSP, which is already fabricated
- To assess the merit of the new techniques using experimental silicon through comparisons of the projected power reduction (in simulation) and actually measured reduction of new DSP; assessment of the commercial impact

LOVO Low Output VOltage DC/DC converters for low power applications

- Development of technical solutions for the power supplies of advanced low power systems, comprising the following topics
- New methods for synchronous rectification for very low output voltage power converters

PCBIT Low Power ISDN Interface for Portable PC's

- Design of a PC-Card board that implements the PCBIT interface
- Integrate levels 1 and 2 of the communication protocol in a single ASIC
- Incorporate power management techniques in the ASIC design:
  - system level: shutdown of idle modules in the circuit
  - gate level: precomputation, gated-clock FSMs

COLOPODS Design of a Cochlear Hearing Aid Low-Power DSP System

- Selection of a future oriented low-power technology enabling future power reduction through integration of analog modules
- Design of a speech processor IC yielding a power reduction of 90% compared to the 3.3 Volt implementation

The low power design projects have achieved the following results:

- Projects, who have designed a prototype chip, can demonstrate a power reduction of 10 to 30 percent.
- New low power design libraries have been developed.
- New proven low power RF architectures are now available.
- New smaller and lighter mobile equipment is developed.

Instead of running a number of Esprit projects at the same time independently of each other, during this pilot action the projects have collaborated strongly. This is achieved mostly by the novelty of this action, which is the presence and role of the coordinator: DIMES - the Delft Institute of Microelectronics and Submicron-technology, located in Delft, the Netherlands (http://www.dimes.tudelft.nl). The task of the coordinator is to co-ordinate, facilitate, and organize:

- The information exchange between projects.
- The systematic documentation of methods and experiences.
- The publication and the wider dissemination to the public.

The most important achievements, credited to the presence of the coordinator are:

- New personnel contacts have been made, and as a consequence the resulting synergy between partners resulted in better and faster developments.
- The organization of low power design workshops, special sessions at conferences, and a low power design web site, http://www.esdlpd.dimes.tudelft.nl. At this site all public reports of the projects can be found and all kind of information about the initiative itself.
- The used design methodology, design methods and/or design experience are disclosed, are well documented and available.

Based on the work of the projects, in cooperation with the projects, the publication of a low power design book series is planned. Written by members of the projects this series of books on low power design will disseminate novel design methodologies and design experiences, which were obtained during the runtime of the European Low Power Initiative for Electronic System Design, to the general public.

In conclusion, the major contribution of this project cluster is that, except the already mentioned technical achievements, the introduction of novel knowledge on low power design methods into the mainstream development processes is accelerated.

We would like to thank all project partners from all the different companies and organizations who make the Low Power Initiative a success.

Rene van Leuken, Reinder Nouta, Alexander de Graaf
Delft, July 2002

TABLE OF CONTENTS:

List of Figures

LIST OF TABLES

CHAPTER 1

INTRODUCTION

1.1 Motivation

The microelectronics has been one of the most quickly developing fields in the past few decades. In the legendary publication of Dr. Gordon E. Moore [1], he predicted that the number of the transistors per integrated circuits would double every 18 months. Fig. 1.1 demonstrates that the Moore's law is still valid even after 35 years. With the increasing integration density, the functional density also increases. Consequently, the demand to build single-chip systems, which previously consisted of several separated chips, is growing. This is so called "system-on-a-chip" or "system-on-silicon", abbreviated as SoC and SoS, respectively. The system-on-a-chip has not only advantage of reduction of the system volume. Its power dissipation per area unit also decreases. Besides, the number of the corresponding parasitic components is also reduced. This leads to possibilities to operate systems at higher speed [2] and improves reliability.

Physical quantities occurring in nature are intrinsically analog signals. However, digital signals can be stored, transmitted, and used in computations easier than analog signals. Hence, mixing of the analog and the digital signals cannot be avoided. Among all devices, the *analog-to-digital converter* (*A/D converter* or *ADC*) and the *digital-to-analog converter* (*D/A converter* or *DAC*) are the two bridges between the analog and the digital worlds. The classical analog integrated circuits employed bipolar transistors, which yielded lower *equivalent noise voltage* than their CMOS counterpart [3]. But, since the CMOS digital circuits are smaller and dissipate less power dissipation than the bipolar circuits, the CMOS analog circuits have become mainstream of the analog circuit technology in the past decade.

Nevertheless, the system-on-a-chip also increases the complexity of the chip and, as a consequence, the design effort. Design methodologies to im-

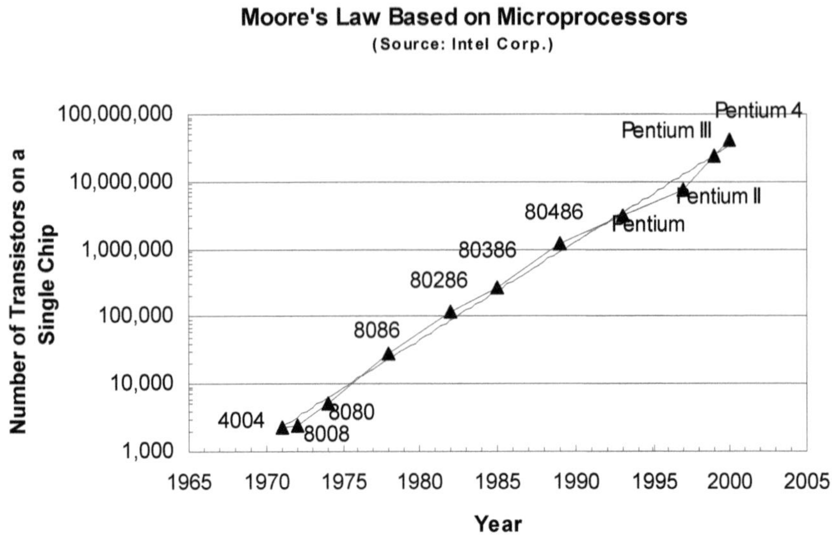

Fig. 1.1: Demonstration of Moore's law and the development of microprocessors

prove the reusability of the single functional blocks are essential to trim down the design costs. As a result, modularization of the analog functional blocks and methods to predict chip parameters, especially the power dissipation and the die area, are needed. A fully described analog functional block is not only reducing the design effort, but it is also part of the "silicon intellectual property" (SIP) of the whole integrated system.

This book is devoted to the design and optimization of high-speed A/D converters. Traditional applications of high-speed A/D converters include the instrumentation and signal processing, e.g. for professional and consumer video, radar warning and guidance systems, spectrum analyzers, medical imaging, and sonar. More advanced application include, for example, digital communications, digital measurement techniques, high definition TV (*HDTV*), and set-top boxes for direct broadcast satellites (*DBS*), which need dual 6~8-bit high-speed A/D converters to perform *quadrature demodulation* [4]. For the last application, good matching between these two A/D converters is necessary, apart from their high-speed performance.

Aim of this book is to develop a family of CMOS A/D converters, which can yield high performance under low-power and high-speed conditions and be integrated into embedded systems. The resolution covers between

8~10 bit at first and extends to 12 bit later on. To reduce the effort of integrating these A/D converters to various applications, components of this A/D converter have to be built like the well-known *"LEGO brick"*, i.e. modular design of the components plays an important role.

1.2 Organization of the Book

In Chapter 2, the principles of the analog-to-digital conversion will be discussed. Main criteria of the A/D converter are given and several important high-speed architectures will be introduced. The components of the CMOS folding and interpolating A/D converter will be treated in detail in Chapter 3. Chapter 4 is devoted to the implementation of the prototype and its corresponding modeling and Chapter 5 to the physical design of low-power, high-speed embedded CMOS A/D converter. Chapter 6 contains a discussion of the evaluation and measurement of the implemented folding and interpolating A/D converter. Chapter 7 gives general conclusion and outlooks of this book.

CHAPTER 2

ANALOG-TO-DIGITAL CONVERSION

Epitome

2.1 Concepts of A/D Conversion and D/A Conversion

2.1.1 Classification and Transformations of Signals

Physical quantities that can be converted into electrical signals, for example sound, light, temperature, and magnetic field, yield intrinsically analog signals. Due to the convenience of processing and storage of the digital signals, conversion of such analog signals into digital signals becomes an important challenge of today's microelectronics, especially conversion of high-speed signals, such as light, radar, and ultrasound signals, into digitized signals for the subsequent signal processing.

Generally speaking, the signal types can be classified into four categories, as illustrated in Fig. 2.1 [5]. The x-axis corresponds to the time t and can exhibit either continuous or discrete values, i.e., it is either nonquantized or

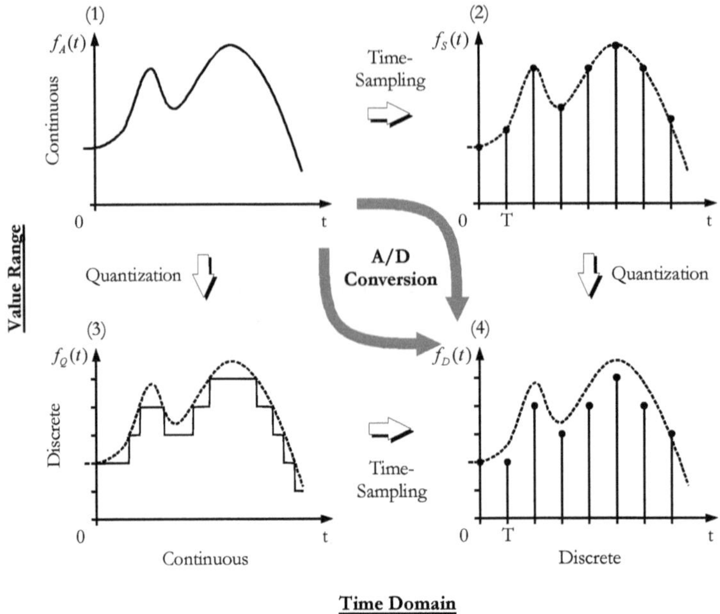

Fig. 2.1: Classification of signals

quantized in time, respectively. The y-axis carries the signal magnitude $f(t)$ and can also exhibit continuous or discrete values. The category (1) contains signals of continuous values with respect to time and magnitude. Using the *analog time-sampling operation*, signals of the category (1) are transformed into the category (2), which now contains only discrete-time values.

Furthermore, quantizing the signal magnitude of the signals in the category (2) leads to the signal category (4), which contains signals quantized in magnitude and time. The signals in the category (4) are already digitized signals, since both the x-axis and the y-axis exhibit discrete values. The converting process from the category (1) via the category (2) to the signal category (4) is one type of *analog-to-digital conversion* (*A/D conversion*).

The typical feature of this kind of A/D conversion is that the sampling processing is carried out using analog circuits. Typical architectures of such high-speed A/D conversion include the *two-step flash ADC*, the *subranging ADC*, the *pipeline ADC*, the *time interleaved ADC*, and the *successive approximation register ADC* (*SAR ADC*). The *flash ADC* and the *folding and interpolating ADC* can also use an analog time-sampling device as their *analog front-end* (*AFE*). But these two converters do not require necessarily such a device for their operation and are of more importance for the A/D conversion with

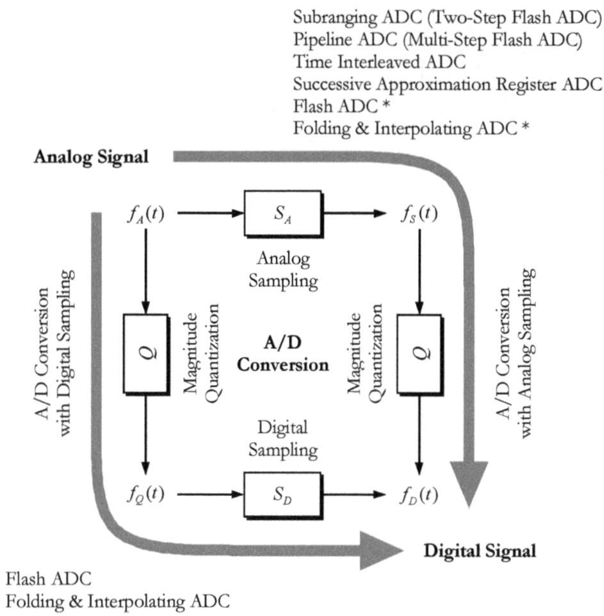

Fig. 2.2: Classification of A/D conversion

simple *digital sampling* discussed below. Therefore, these two converter types are marked by an "*" in Fig. 2.2.

Another way of the A/D conversion is to quantize the analog signals of the category (1) with respect to their magnitude, i.e. before the *time-sampling* process is executed. The analog signals are then transformed into signals of discrete values with respect to their magnitude, but remain of continuous-time type as shown in the category (3) in Fig. 2.1. Simply adding a clock signal to the output control unit of the signal category (3) can digitize the variable t of the magnitude-quantized signal of the category (3). The feature of this kind of A/D conversion is that the time-sampling is carried out after the *magnitude quantization* process. This kind of sampling processing is easier to be implemented and realized using low-power circuits than the analog sampling processing discussed above. Typical architectures of this A/D conversion are represented by the flash A/D converter and the folding and interpolating A/D converter.

The both in time and magnitude quantized signals can be coded into suitable binary signals for the further digital signal processing as illustrated in Fig. 2.3.

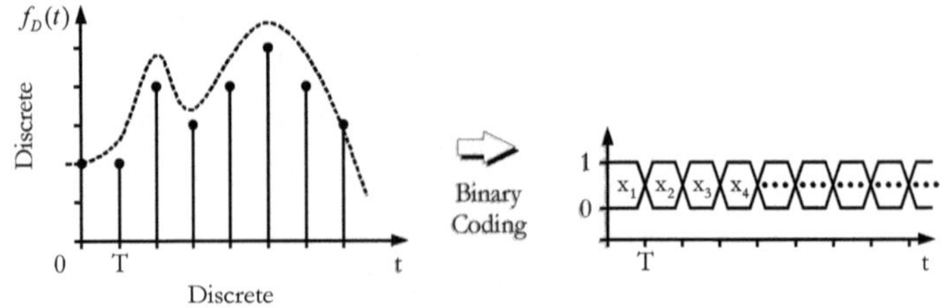

Fig. 2.3: Time schedule of digitized output signals and its binary coding

2.1.2 Principle of A/D Conversion

Summarizing the discussion above it can be stated that there are two different approaches to the analog-to-digital conversion from the viewpoint of signal classification. Fig. 2.2 shows that the main difference of these two conversion approaches is the application of the time-sampling. This is carried out either before or after magnitude quantization, thus yielding the signals $f_S(t)$ and $f_Q(t)$, respectively. In the following, we shall call these two approaches *analog sampling* and *digital sampling*, respectively. Due to the different sampling methods, the conversion errors caused by the two conversion methods are dissimilar.

The transform function of the analog sampling $S_A(t)$ can be written as [5]:

$$S_A(t) = \sum_{n=-\infty}^{\infty} \mathrm{rect}\left(\frac{t-nT}{T_0}\right), \tag{2.1}$$

where the variable T is the sampling period of the clock signal and the variable T_0 is the duration of the actual signal acquisition time.

The transform function of the digital sampling $S_D(t)$ is a special case of the analog sampling $S_A(t)$ as the duration of the acquisition time is $T_0 \to 0$: this can be denoted by Dirac impulse as

$$S_D(t) = \sum_{n=-\infty}^{\infty} \delta(t-nT). \tag{2.2}$$

Furthermore, providing that the transform functions $Q(t)$ of the magnitude quantization of the two kinds of conversion are the same, the digital

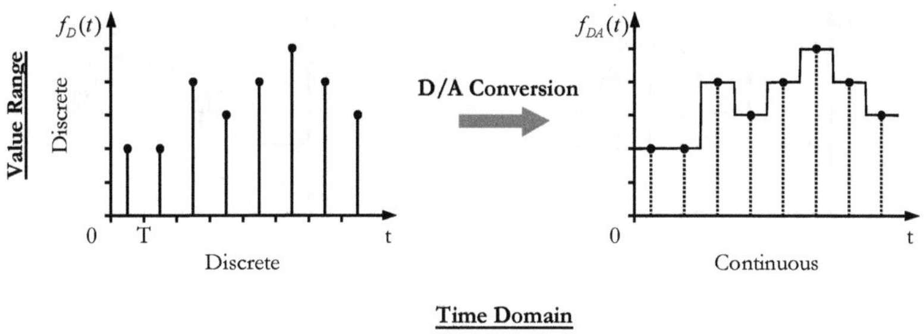

Fig. 2.4: Principle of the D/A conversion

signal $f_{D1}(t)$ converted by using the analog sampling method can be described as:

$$\begin{aligned} f_{D1}(t) &= Q\{f_S(t)\} = Q\{S_A\{f(t)\}\} \\ &= Q\left\{ f(t) * \sum_{n=-\infty}^{\infty} rect\left(\frac{t-nT}{T_0}\right)\right\} \\ &= Q\left\{ f(t) * \left[rect\left(\frac{t}{T_0}\right) * \sum_{n=-\infty}^{\infty} \delta(t-nT)\right]\right\} . \end{aligned} \tag{2.3}$$

There are two kinds of possible errors involved in quantization process. Time-sampling introduces signal aliasing in case of undersampled signals, i.e. where the sampling theorem violated, while magnitude quantization causes a quantization error that cannot be recovered in the ADC output signal. Definition and discussion of the *quantization error* due to the magnitude quantization will be given in Chapter 2.3.1.

The output signal $f_{D2}(t)$ of the A/D conversion based on the digital sampling method can be described as:

$$f_{D2}(t) = S_D\{f_Q(t)\} = S_D\{Q\{f(t)\}\} = Q\{f(t)\} * \sum_{n=-\infty}^{\infty} \delta(t-nT) . \tag{2.4}$$

From Eqs. (2.3) and (2.4) it can be found that the transfer function with the digital sampling function is easier to implement than the analog time-sampling function. The implementation of the analog *sample-and-hold amplifier (SHA)* represents one of the most difficult design tasks in the area of high-speed analog circuits. The high-speed SHA usually contains an *operational transconductance amplifier* (OTA), which consumes much more power

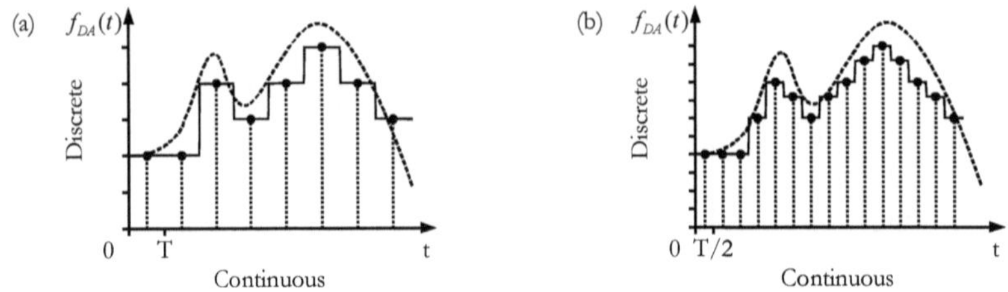

Fig. 2.5: Improvement of the similarity by doubling the clock frequency and resolution

than other devices due to its output stage. This kind of architecture is not suitable for the embedded systems that have severe restrictions on the power dissipation and the die area of the A/D converter.

On the other side, the classical flash A/D converter has limitations due to its architecture since for an n-bit converter it requires 2^n comparators for the magnitude quantization: this imposes a severe limit on resolution for a given power dissipation. As a result, the folding and interpolating A/D converter is a good candidate both for high-speed and low-power applications of analog-to-digital conversion. Detailed discussion concerning investigation of different A/D conversion architectures is presented in the later sections of this chapter.

2.1.3 Principle of D/A Conversion

Since many of the A/D conversion architectures involve internal D/A conversion, for example the pipeline ADC and the SAR ADC, the transfer errors caused by the D/A conversion influence also the performance of the A/D conversion. Unfortunately, the transfer function of D/A conversion is not simply the inverse transfer function $S_D^{-1}(t)$ of the digital sampling processing, despite the signal $f_Q(t)$ in Fig. 2.1 and Fig. 2.2 exhibiting discrete magnitude values in the continuous-time domain and thus appearing similar to the output signal of the D/A converter $f_{DA}(t)$ illustrated in Fig. 2.4.

The relationship between the input signal and the output signal of a D/A converter must be derived from the signal $f_D(t)$ and is thus described as follows:

$$f_{DA}(t) = f_D(t) * \sum_{n=0}^{\infty} \text{rect}\left(\frac{t - nT}{T}\right). \tag{2.5}$$

Even if the D/A converter were ideal and introduced no conversion errors, the output signal of the D/A converter $f_{DA}(t)$ would be related to the original analog signal $f_A(t)$ or the signal $f_Q(t)$ from Fig. 2.1 and Fig. 2.2, but it would be still dissimilar. Improving the clock frequency f_{Clk} and the resolution of the D/A converter can improve the similarity between the signal $f_{DA}(t)$ and the signal $f_A(t)$ and/or the signal $f_Q(t)$ although magnitude quantization errors remain. The quantization errors can be only reduced by employing a higher resolution.

Fig. 2.5 shows the improvement of the similarity between the analog signal $f_A(t)$ and the output signal of the D/A converter $f_{DA}(t)$ by doubling the clock frequency f_{Clk} and the resolution of the D/A converter. Due to the increased clock frequency and resolution, the additional digital data can be obtained using digital interpolation to reduce the conversion errors.

2.2 Sampling Theorem of the A/D Conversion

2.2.1 Nyquist Frequency

Beside the magnitude quantization process, the time-sampling process represents the most important procedure of the analog-to-digital conversion for both the analog sampling and digital sampling. The sampling process can be observed in the time domain as well as in the frequency domain. For A/D conversion used in traditional instrumentation and signal processing applications, the time-domain performance is more significant than the frequency-domain performance [4]. Since applications of A/D conversion in communications have been growing very fast in the past decade, comprehensive investigations emphasizing the frequency-domain analysis have been gaining recently more importance.

One of the most important theorems concerning the sampling process is called the *Nyquist criterion*. It relies on the definition of so called *Nyquist frequency*. For its definition the frequency range between zero and half of the clock frequency f_{Clk} used for time-sampling is defined as the 1st *Nyquist zone* (NZ) and the frequency range between half of the clock frequency and the

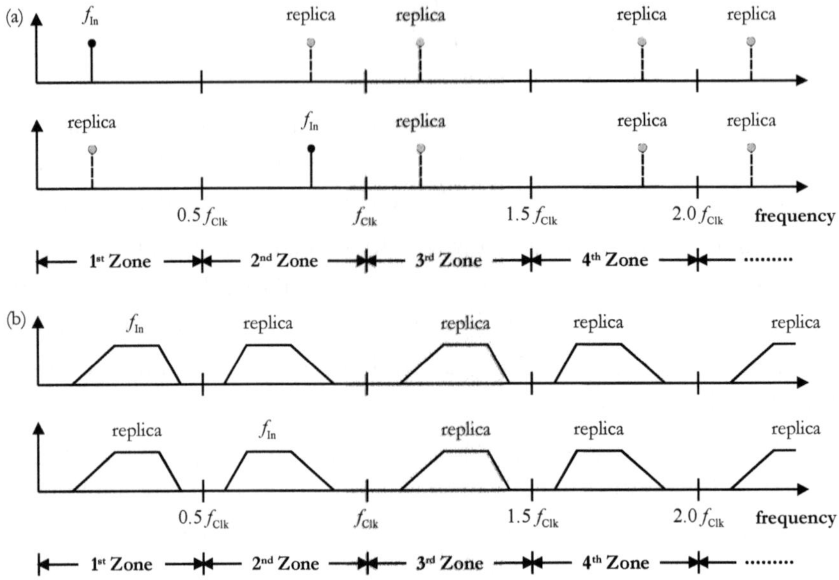

Fig. 2.6: Relationship between the frequencies of the input signal and its replicas

clock frequency f_{Clk} is defined as the 2nd Nyquist zone as shown in Fig. 2.6. A general form of the ith Nyquist zone can be written as:

$$(i-1)\cdot\frac{f_{Clk}}{2} \leq i^{th} \text{ Nyquist Zone} < i\cdot\frac{f_{Clk}}{2} \quad . \tag{2.6}$$

To categorize the time-sampling processes in terms of Nyquist zones, the definition concerning the *oversampling*, the *undersampling*, and the *Nyquist sampling* can be summarized as followed:

> If and only if the frequency of the input signal falls inside the 1st Nyquist zone, the sampling process can be designated as oversampling. Otherwise, if the input signal is outside of the 1st Nyquist zone, the corresponding sampling process corresponds to undersampling. Sampling process, where the input signal is just half of the clock frequency, is categorized as Nyquist sampling.

Fig. 2.7 shows a sine function $f_{In}(t)$ in the time domain which is sampled at three different sampling periods, T_{Clk1}, T_{Clk2}, and T_{Clk3}. The sampling period T_{Clk2} is exactly half of the period of the sine function $f_{In}(t)$. That is to

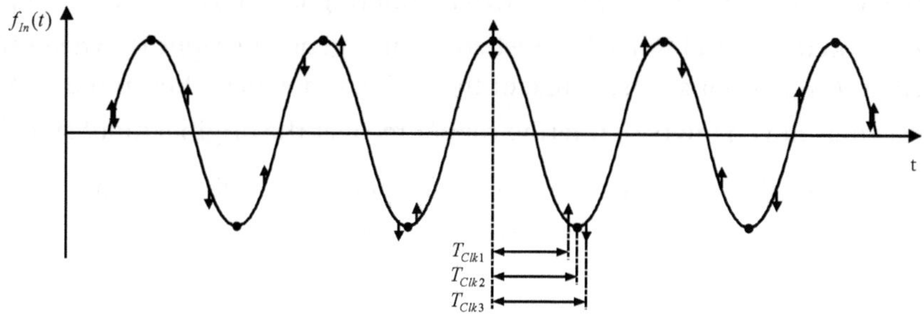

Fig. 2.7: Nyquist sampling, undersampling, and oversampling

say, every sine wave is sampled in this case at two time instants within its period. For the convenience of discussion, the sampling instants T_{Clk2} indicated by the dots in Fig. 2.7 have been chosen at the peaks of the sine wave. T_{Clk1} and T_{Clk3} have been chosen slightly shorter and longer than T_{Clk2}, respectively. The relationship of their frequencies can be written as follows:

$$f_{Clk2} = 2 f_{In} \quad \text{or} \quad f_{In} = \frac{f_{Clk2}}{2} \,, \tag{2.7}$$

$$f_{Clk2} < f_{Clk1} \quad \text{or} \quad f_{In} < \frac{f_{Clk1}}{2} \,, \tag{2.8}$$

and

$$f_{Clk2} > f_{Clk3} \quad \text{or} \quad f_{In} > \frac{f_{Clk3}}{2} \,. \tag{2.9}$$

In the first case, i.e. given by Eq. (2.7), the clock frequency f_{Clk2} is just twice the input frequency f_{In}. This clock frequency f_{Clk2} is defined as the *threshold frequency* of the Nyquist sampling. This threshold frequency of the Nyquist sampling defines the *Nyquist frequency* f_N, which can be written as:

$$f_N \equiv \frac{f_{Clk2}}{2} = f_{In} \,. \tag{2.10}$$

For all signals with frequencies below or at f_N the entire information contained in the original signal $f_{In}(t)$ is uniquely defined in the 1st Nyquist zone, since all signal replicas generated by the time sampling fall into frequency bands above f_N (see Fig. 2.6). Hence the information of the input signal can be fully recovered by lowpass filtering up to f_N. Note that input signals occurring in higher Nyquist zones also generate replicas, when sampled at f_{Clk2}. Among others, one of the replicas, however, lies in the 1st Nyquist zone: now a lowpass filtering cannot recover the original input signal.

Given that the input signal $f_{In}(t)$ is not composed of single frequency f_{In}, the Nyquist frequency f_N can then be defined as:

$$f_N \equiv f_C \,, \tag{2.11}$$

where the frequency f_C is the maximum frequency of the input signal $f_{In}(t)$.

For the second situation, the frequency f_{In} of the sine wave at the input is still lower than half of the clock frequency f_{Clk1}. The coarse waveform of the input sine function can also be preserved in the sampled data. This kind of sampling process is called oversampling because the sampling frequency is greater than twice the input frequency. Since the frequency of the input signal is lower than half of the sampling frequency, it is also called *baseband sampling*.

But, if the input frequency f_{In} is higher than half of the clock frequency f_{Clk3} as described in Eq. (2.9), information in the input cannot be uniquely represented. In this case, it is an undersampling situation. Contrary to the baseband sampling discussed above, this kind of sampling process is also known as *harmonic sampling, bandpass sampling, IF sampling,* and *direct IF to digital conversion* in the literature [4, 6].

2.2.2 Aliased Signals

Fig. 2.6 shows also the latent ambiguity of the sampled signal in the frequency domain, namely the *aliases* (also *replicas* or *images*) of the original signal f_{In}. The aliased signals occur around every multiple of the sampling frequency f_{Clk}. Exact description of the positioning of signal components can be written as follows [4]:

$$\left|\pm i \cdot f_{Clk} \pm f_{In}\right| \,, \text{ where } i = 1,2,3\ldots\,. \tag{2.12}$$

These aliased signals make it impossible, for example, to distinguish between input signals inside the 1st Nyquist zone and signals outside the 1st

Nyquist zone as illustrated in Fig. 2.6(a). In certain circumstances, the oversampling process can have the same behavior as the undersampling process.

In addition, the performance of the sampling process is degraded not only by unexpected signals near the input signal f_{In} inside the same Nyquist zone, but also by the signals which have image components falling near to the input signal f_{In}. These aliased signals produce *spurious frequency components*, which decrease the signal-to-distortion ratio of the sampled signal in the same Nyquist zone of the input signal f_{In}.

An *anti-aliasing filter* can reduce this kind of disturbance. For the baseband sampling, the performance of the anti-aliasing filter depends on the distance between the Nyquist frequency f_N, the corner frequency, and the stopband frequency of the filter and the amount of the required signal attenuation.

For the harmonic sampling, to ensure the carrier frequency f_C of the input signal is positioned in the center of a Nyquist zone, there are two conditions to be taken into consideration [4]:

$$f_{Clk} > 2 \cdot \Delta f_{In} \,, \tag{2.13}$$

and

$$f_{Clk} = \frac{4 \cdot f_C}{2 \cdot NZ - 1} \,, \; NZ = 1,\ 2,\ 3,\ \dots \,, \tag{2.14}$$

where Δf_{In} represents the bandwidth of the input signal f_{In} and NZ indicates the number of the Nyquist zone in which the carrier frequency f_C is to be found.

2.3 Quantization Process of the A/D conversion

2.3.1 Quantization Error

The *quantization error* is the nature of the conversion process itself. This kind of error is irreversible and cannot be corrected [7]. Fig. 2.8 shows the transfer functions of an A/D converter.

The diagonal line Γ describes the ideal analog-to-digital transfer function of an ideal A/D converter exhibiting an infinite resolution and the *least significant bit* (*LSB*) of this transfer function is infinitesimal. The stair-like characteristic curve D describes the theoretical transfer function of a 3-bit A/D

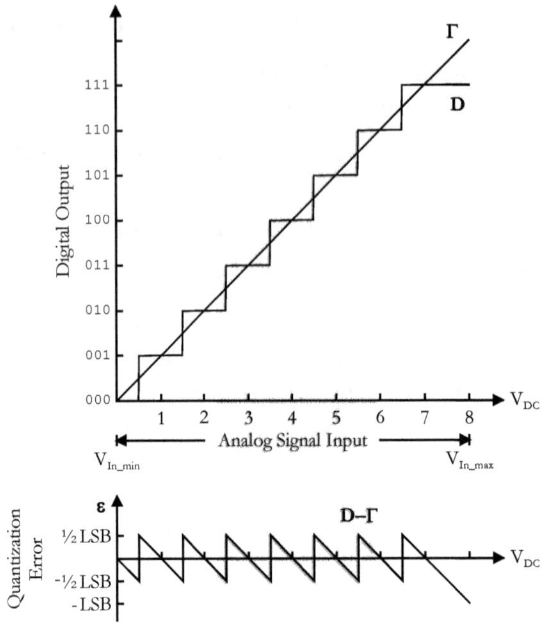

Fig. 2.8: Quantization error of a 3-bit A/D conversion

conversion. The difference between the ideal transfer function of an A/D converter with infinite resolution and the theoretical transfer function of A/D converter with finite resolution is designated as $D-\Gamma$, as illustrated in Fig. 2.8, and is defined as the quantization error ε of A/D conversion. Since the quantization error ε can be understood as random signal when performing the *discrete Fourier transform* (*DFT*) analysis of the output signal, the quantization error ε is also called *quantization noise* in the literature. According the arrangement in Fig. 2.8, the quantization error ε can be described as follows:

$$-\tfrac{3}{2}\,LSB \le \varepsilon \le \tfrac{1}{2}\,LSB\,. \tag{2.15}$$

The definition of the transfer function D as shown above leads to an unsymmetrical quantization error ε in Eq. (2.15). As a consequence, the integration of the quantization error ε over the analog input signal range equals to $-\tfrac{1}{2}LSB^2$, which is not equal to zero. This makes the noise characterization of the quantization error ε more difficult due to nonzero mean value.

An alternative definition is to shift both the ideal transfer function Γ and the theoretical transfer function D of the A/D converter by a ½ LSB to the

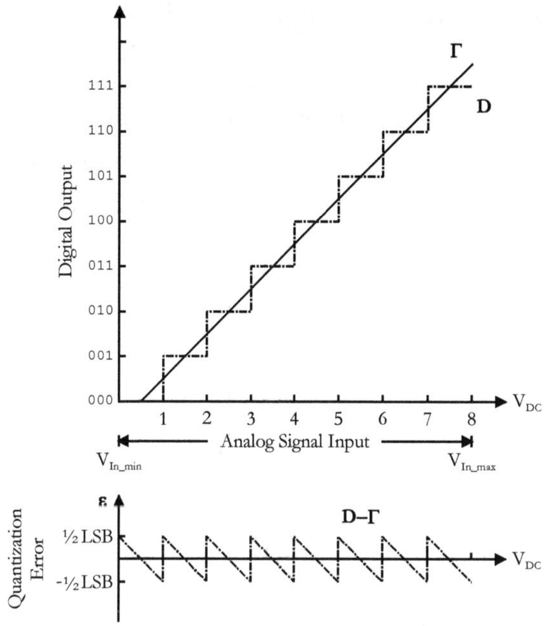

Fig. 2.9: Shifted transfer function and the corresponding quantization error

right, as depicted in Fig. 2.9. This leads to a symmetrical distribution of quantization error ε, which can be described as:

$$-\tfrac{1}{2}\ LSB \le \varepsilon \le \tfrac{1}{2}\ LSB\ . \tag{2.16}$$

As a consequence of the definition above and Fig. 2.9, the integration of the quantization error ε over the input signal range equals to zero.

2.3.2 Estimate of Theoretical Quantization Error

To estimate the value of the quantization error ε methods like the *mean square error* and the *root-mean-square error* (*RMS error*) are usually employed in the open literature. This allows the computation of the absolute value of the error without canceling its positive values by negative ones. The definition of the mean square error can be described as follows [8, 9]:

$$Mean\ Square\ Error \equiv \frac{1}{b-a} \cdot \int_a^b [g(x) - f(x)]^2\, dx\ , \tag{2.17}$$

where the function $g(x)$ is the approximation function of the function $f(x)$ and the interval $[a,b]$ indicates where the approximation has been applied. Applying Eq. (2.17) to the A/D conversion shown in Fig. 2.9 leads to the following relationship:

$$Mean\ Square\ Error\ \overline{\varepsilon^2} = \frac{1}{LSB} \int_{-\frac{1}{2}LSB}^{\frac{1}{2}LSB} \varepsilon^2 d\varepsilon = \frac{1}{12} \cdot LSB^2\ . \tag{2.18}$$

The corresponding root-mean-square value of the quantization error ε_{RMS} can be written as:

$$\varepsilon_{RMS} \equiv \sqrt{\overline{\varepsilon^2}} = \frac{1}{2\sqrt{3}} \cdot LSB\ . \tag{2.19}$$

2.4 Performance of A/D Converters

2.4.1 Differential Nonlinearity and Integral Nonlinearity

The nonlinearity describes basically a static performance of an A/D converter. There are two different types of nonlinearity: differential and integral. The *differential nonlinearity* (DNL) describes the maximum difference between the theoretical and real transfer functions of an A/D converter measured between adjacent LSB values over the full conversion range. Theoretically each transition is equal to 1LSB, which corresponds to a zero differential nonlinearity.

The DNL value of a particular digital output signal *j* of the A/D converter can be written as:

$$DNL_j \equiv Width_{LSB} - Width_{Output,j}\ . \tag{2.20}$$

Assuming that the width of the digital output signal *j* is equal to a single LSB, the DNL value is zero. In contrast, if there is any missing code in the conversion process, then the width of this missing code is zero and the DNL value for this missing code will be 1 *LSB*.

The *integral nonlinearity* (*INL*) indicates the maximum deviation of the real transfer function (i.e. of the realized A/D converter) from its ideal transfer function with gain and offset errors nulled to zero.

The relationship between the differential nonlinearity and the integral nonlinearity can be described by the following equations:

$$INL_j = \sum_{k=0}^{k=j} DNL_k \ , \tag{2.21}$$

and

$$DNL_j = INL_{j+1} - INL_j \ . \tag{2.22}$$

2.4.2 Signal-to-Noise Ratio

The *signal-to-noise ratio* (*SNR*) is one of the most important parameters to evaluate the dynamic performance of the A/D converter. It describes the resolution of the A/D converter in dynamic range.

Because the quantization error ε represents an innate feature of the A/D conversion and cannot be eliminated, the theoretical value of the SNR is given by the quantization error ε, providing that there is no external noise, which can also affect the conversion process. For a converter exhibiting *n*-bit resolution, the peak-to-peak value of the full-scale input signal A_{PP} can be written as [7, 10]:

$$A_{PP} = 2^n \cdot LSB \ . \tag{2.23}$$

Hence, if the input signal is a sine wave, the RMS value of the input signal can be expressed as:

$$A_{RMS} = \frac{A_{PP}}{2} \cdot \frac{1}{\sqrt{2}} = \frac{2^{n-1} \cdot LSB}{\sqrt{2}} \ . \tag{2.24}$$

As stated above, for the theoretical computation of the SNR value, no other noise source is to be considered, except the quantization error ε. The theoretical SNR can be thus described as:

$$SNR_{Theoretical} = \frac{A_{RMS}^2}{\varepsilon_{RMS}^2} = \left(\frac{2^{n-1} \times LSB}{\sqrt{2}} \times \frac{2\sqrt{3}}{LSB} \right)^2 = \left(2^n \times \sqrt{1.5} \right)^2 \ . \tag{2.25}$$

In this way, the theoretical value of SNR in decibel expression can be written as:

$$\begin{aligned} SNR_{Theoretical} &= 20 \cdot \log\left(2^n \cdot \sqrt{1.5}\right) \\ &= 6.02n + 1.76 \ [\text{dB}] , \end{aligned} \tag{2.26}$$

where n denotes the resolution of the converter. The 1st term is related to the resolution of the converter directly. In contrast, the 2nd term 1.76 dB is contribution of the signal waveform: in this case the input signal waveform is a sine wave and the quantization error is a kind of saw tooth function.

Providing that the SNR value is measured by the spectrum analyzer, the *effective number of bits* (*ENOB*) can be deduced by solving the variable n of Eq. (2.26):

$$ENOB = \frac{SNR_{Real} - 1.76}{6.02} \ [\text{dB}] . \tag{2.27}$$

This equation is only valid if the noise is measured over the entire *Nyquist bandwidth* from DC to $f_{Clk}/2$. Providing that the considered signal bandwidth f_{BW} is lower than $f_{Clk}/2$, the corresponding noise in the SNR calculation is lower due to lower bandwidth. The measured value of the SNR then improves by a factor of $f_{Clk}/(2 \cdot f_{BW})$. The relationship between the SNR_{Real} and the ENOB can be supplemented as:

$$SNR_{Real} = 6.02 \cdot ENOB + 1.76 + 10 \cdot \log\left(\frac{f_{Clk}}{2 \cdot f_{BW}}\right) \ [\text{dB}] . \tag{2.28}$$

The last term is also named as *processing gain* or *FFT processing gain* in the literature.

The parameter ENOB represents a comprehensive description of the A/D converter. When measured at real converters, it covers all possible errors of the A/D conversion, including the dynamic DNL and INL errors, *missing codes*, *clock jitter*, and other noises. The theoretical value of the parameter ENOB represents the maximum possible resolution, but it is impossible to reach in practice.

2.4.3 Dynamic Range

The *dynamic range* (*DR*) is defined as the ratio between the largest input signal and the smallest representable signals, namely the LSB. This ratio expresses the number of the representable states of the A/D converter. An analog-to-digital converter with n-bit resolution has dynamic range of 2^n. In case that the dynamic range is written in decibel form, the description of the dynamic range is as follows:

$$DR \equiv \frac{V_{In\_max} - V_{In\_min}}{LSB} = 2^n = 20 \cdot \log\left(2^n\right)\ [\text{dB}] = 6.02n\ [\text{dB}], \tag{2.29}$$

which is equal to the 1st term of Eq. (2.26). As a consequence, Eq. (2.26) can be written for sine wave signals as:

$$SNR_{Ideal} = DR + 1.76\ [\text{dB}]. \tag{2.30}$$

2.4.4 Signal-to-Noise and Distortion Ratio

Signal-to-noise and distortion ratio (*SINAD, SNDR,* or *S/N+D*) is defined as ratio of the root-mean-square (RMS) value of the input signal f_{In} to the mean value of the *root-sum-squares* (*RSS*) of all other spectral components, including all harmonics, but excluding the DC component [4].

The difference between SINAD and SNR is that SINAD includes spectral components of all harmonics, while the first 5 harmonics, which are generally the dominant harmonics, are not considered in SNR. Hence, measured values of SNR are higher than measured values of SINAD. In the literature, the common definition of the performance ENOB is not based on SNR, as shown in Eq. (2.27), but it is based on SINAD. In this case Eq. (2.27) can be rewritten as:

$$ENOB = \frac{SINAD_{Real} - 1.76}{6.02}\ [\text{dB}]. \tag{2.31}$$

SINAD is normally plotted as a function of the input frequency f_{In} because it includes all spectral components, which make up noise (including thermal noise) and distortion. Often, it is also plotted for various input amplitudes [4].

Table 2.1: Critical parameters of A/D converters

Typical Applications	Performance Issues	Critical ADC Parameters																		
		Bandwidth	DNL	INL	ENOB	SINAD	SFDR	THD	IMD	NPR	Noise-to-distortion ratio	Out-of-range recovery	Fullscale step response	Bit Error Rate	Word Error Rate	Short-term settling	Long-term stability	Differential Gain Error	Differential Phase Error	Monotonicity
Audio	Power consumption. Crosstalk and gain matching.					X		X												
Automatic control	Transfer function. Crosstalk and gain matching. Temperature stability.															X	X			X
Digital oscilloscope Waveform recorder	SINAD for wide bandwidth amplitude resolution. Low thermal noise for repeatability.	X			X	X						X		X	X					
Geophysical	Miliherty response.					X		X									X			
Image processing	DNL for sharp-edge detection. High-resolution at switching rate. Recovery for blooming.		X	X	X	X						X	X							
Radar and sonar	SINAD and IMD for clutter cancellation and Doppler processing.				X	X	X		X			X								
Spectrum analysis	SINAD and SFDR for high linear dynamic range measurements.				X	X	X													
Spread spectrum communication	IMD for quantization of small signals in a strong interference environment. SFDR for spatial filtering. NPR for interchannel crosstalk.				X	X	X		X	X	X									
Telecommunication Personal communications	Wide input bandwidth channel bank. Interchannel crosstalk. Compression. Power consumption.					X	X		X	X				X	X					
Video	Differential gain and phase errors. Frequency response.		X			X	X											X	X	
Wideband digital receivers Signal intelligence Electronic intelligence Communications intelligence	Linear dynamic range for detection of low-level signals in a strong interference environment. Sampling frequency.					X	X		X											

2.4.5 Total Harmonic Distortion

Total harmonic distortion (*THD*) is defined as the ratio of the root-mean-square (RMS) value of the input signal f_{In}, i.e. the fundamental signal, to the mean value of the root-sum-square (RSS) of its harmonics. Generally, only the first 5 harmonics are significant and dominate the result.

Moreover, *total harmonic distortion plus noise* (*THD+N*) considers not only the spectral components of the harmonics, but also all noise components (excluding the DC component) inside the specific bandwidth. Given that the specific bandwidth is the complete 1st Nyquist zone, i.e. from DC to $f_{Clk}/2$, values of the THD+N are equal to values of SINAD.

2.4.6 Spurious Free Dynamic Range

The performance of *spurious free dynamic range* (*SFDR*) of an A/D converter is defined as the ratio of the root-mean-square (RMS) value of the input signal f_{In} to the RMS value of the peak spurious spectral content, which is measured over the entire first Nyquist zone, i.e. from DC to $f_{Clk}/2$. The SFDR is generally plotted as a function of the input signal amplitude and expressed relatively to the signal amplitude (*decibels below carrier, dBc*) or to the full scale of the A/D converter (*decibels below full scale, dBFS*) [4].

If the amplitude of the input signal f_{In} is near the full scale of the A/D converter, the peak spectral spur is generally dominated by one of the first few harmonics of the fundamental frequency. But, if the signal falls several dB below the full scale of the A/D converter, other spurs, which are not direct harmonics of the input signal f_{In}, generally occur. This is because of the differential nonlinearity of the transfer function of the A/D converter itself. Hence, the SFDR considers all sources of distortion disregarding their origin [4]. The SFDR is one of the most significant specifications for an ADC used in a communication applications.

Due to the variety of applications of A/D converters, it is impossible to make a comprehensive listing comparing all parameters of A/D converters and their corresponding applications. Table 2.1, published by IEEE Standards Department [11], shows some critical parameters of A/D converters in respect to many of the most common applications of A/D converters.

2.5 Topologies and Architectures of High-Speed A/D Converters

Generally speaking, high-speed A/D converters are A/D converters with sampling frequencies f_{Clk} exceeding $1\ \text{MHz}$. The high-speed A/D converter

Table 2.2: Classification of high-speed A/D converters

Direct A/D converters	
Without feedback	With feedback
Flash ADC	
Two-step flash ADC	
Subranging ADC	
Multi-step flash ADC	
Pipeline ADC	
Time-interleaved ADC	
	Successive approximation register ADC
Folding and interpolating ADC	

is categorized as *"direct A/D converter"* in the literature [12]. Apart from the sampling frequency f_{Clk}, the input frequency f_{In} of the intended application and the required resolution are also of main concern and have great influence on the A/D converter architecture to be chosen. In this section, topologies and architectures of high-speed A/D converters will be introduced and compared.

Before we begin this investigation, we have to clarify the definitions of *"resolution"*, *"accuracy"*, and *"precision"* at first.

Resolution specifies the theoretical number of presentable states, which the analog input signal f_{In} can be resolved into. The concepts of "accuracy" and "precision" are easily confused. Accuracy designates how close the real A/D converter comes to the theoretical value. The accuracy can be explored in the time domain, where specifications are static, such as DNL, INL, *offset error*, and *gain error* [13]. On the other side, precision means the repeatability of the A/D converter. A precise system is not necessarily an accurate system.

Survey of the state-of-the-art A/D converters shows that the high-speed A/D converters can be classified into two groups, i.e. with feedback or without feedback (see Table 2.2). The architectures without feedback have more advantages concerning to the high-speed conversion. Details of each architecture will be shown in the following.

2.5.1 Flash A/D Converter

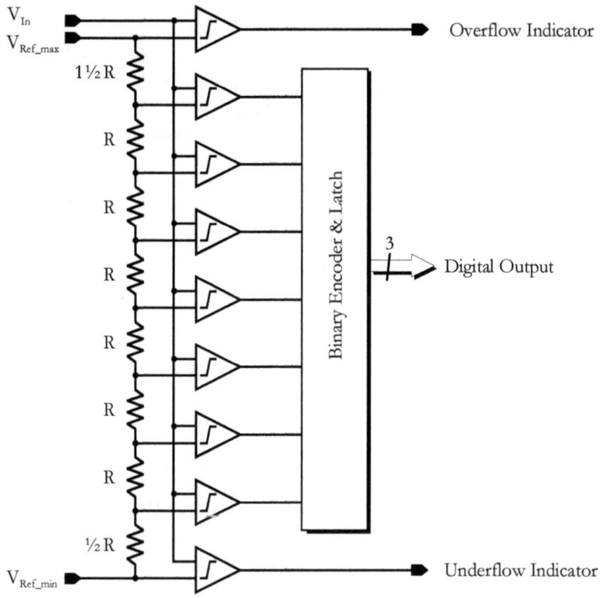

Fig. 2.10: Block diagram of a 3-bit flash A/D converter

The well-known *flash A/D converter* represents the classical high-speed A/D converter architecture [4, 10, 14-17]. Recent research results show that the flash A/D converters featuring 6-bit resolution and fabricated in $0.35\ \mu m$ – CMOS technology, achieve 1.3 GSPS with 500 mW power dissipation at 3.3 V power supply voltage [18] or 1.1 GSPS with 300 mW also at 3.3 V [19]. An embedded implementation [20] for a mixed-signal single chip fabricated in $0.18\ \mu m$ – CMOS technology for DVD systems contains a 7-bit flash ADC, a 32-bit RISC CPU, servo DSP and 16 Mb DRAM.

A block diagram of a 3-bit flash A/D converter is illustrated in Fig. 2.10. The main components of a flash A/D converter are comparators and related reference voltages, which can be obtained using a combination of reference voltage sources and a resistor string. The digital output of the comparator is encoded using digital logic. The input signal range of the converter, $V_{In\_\min}$ and $V_{In\_\max}$, is defined by the reference voltage sources, $V_{Ref\_\min}$ and $V_{Ref\_\max}$. The flash A/D converter is also called *parallel A/D converter* because all comparators are connected in parallel [4].

In case that the input signal is positioned between the kth and $(k+1)^{th}$ reference voltages, the comparators with reference voltage below the input signal level yield a logical "1" at their outputs. On the other hand, the com-

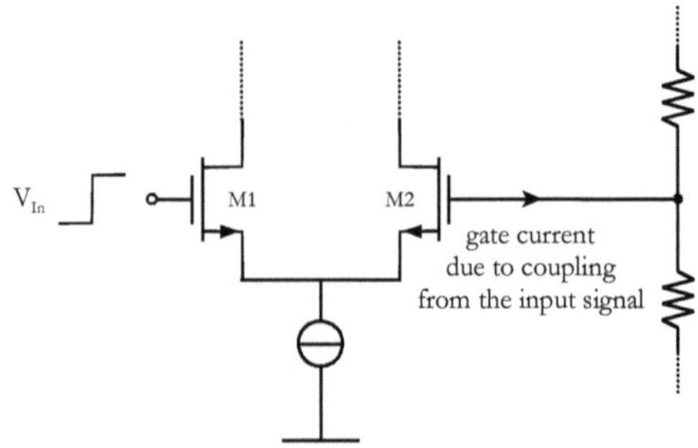

Fig. 2.11: Current disturbance of the resistor ladder

parators with reference voltage above the input signal level yield a logical "0" at their outputs. The behavior of the output signals of the comparators is similar to a mercury thermometer, thus these kinds of output signals are named as *thermometer codes* in the literature [4].

Supposing that the flash A/D converter has function of *out-of-range indication,* thus an n-bit flash A/D converter needs $2^n + 1$ comparators and 2^n resistors, otherwise the flash A/D converter needs only $2^n - 1$ comparators, but it still requires 2^n resistors, as can be seen in Fig. 2.10. This is a very heavy limitation on the flash A/D converter because the total number of the comparators increases exponentially with respect to the resolution; thus the power dissipation and the die area also increase with the same rate.

Since the flash A/D converters are usually employed in high-speed applications, all comparators must be operated at high speed, which is implying high power dissipation. If the resolution is to be increased, the complexity of a single comparator will also increase due to higher performance required. Design techniques, such as *auto-zeroing,* have to be applied to reduce the low-frequency noise and offset problem of the comparators. This is due to the fact that for an n-bit flash A/D converter, all comparators have to reach the n-bit accuracy. This aspect also leads to higher power dissipation and larger die area. Practically, the flash architecture can reach resolutions up to 10-bit, but most of commercially available flash A/D converters exhibit 8-bit.

The next concern is the *input impedance* of the converter. Since all comparators are connected in parallel, the capacitive input impedance increases proportionally with the total number of the comparators. Furthermore, the input impedance of a single comparator can be modeled as combination of a constant capacitance and a variable capacitance, which is function of input signal [3, 4]. As a result, the input impedance varies while the input signal

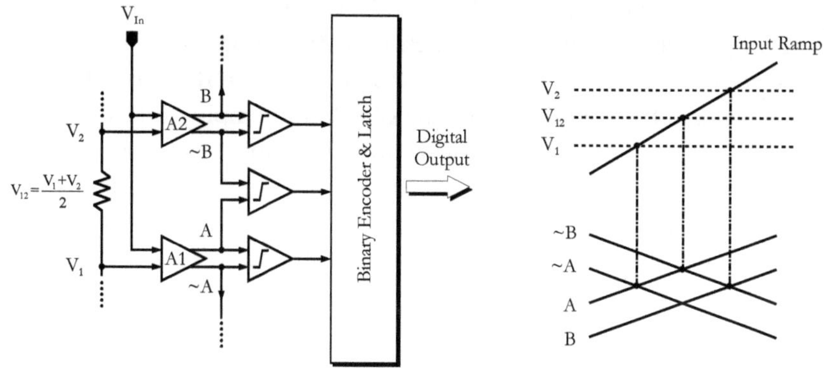

Fig. 2.12: Interpolation design technique for the flash A/D converter

changes from V_{In_min} to V_{In_max}. This causes problems if the input signal is supplied by a nonideal voltage source, because the converter input represents a nonlinear capacitive load for the source. This can cause higher harmonic distortion.

While one input of each comparator is connected to the input terminal of the A/D converter, the other input of the comparator is connected to a resistor string, which generates reference voltages of the converter as shown in Fig. 2.11. We have to consider that there are limits on the resistor size in the resistor string: low resistor value causes high power dissipation and heavy loading of the reference voltage source but it enables fast settling in cases of disturbance; high resistor value requires large die area for the string. Also, high value resistor in the string increases the sensitivity to coupling from the input transients in the comparators: this may cause disturbances at the resistor string.

In addition, the power supply and reference voltages also affect the resolution of the flash A/D converter. With the development of integrated circuits towards lower power supply voltages, the input voltage range $|V_{In\_max} - V_{In\_min}|$ also shrinks rapidly. For example, a 10-bit flash A/D converter needs 2^{10} resistors. In case that the power supply voltage is $3.3\ \text{V}$, the value of the least significant bit (LSB) can be calculated as follows:

$$LSB = \frac{|V_{In\_max} - V_{In\_min}|}{2^{10}} \le \frac{3.3\ \text{V}}{2^{10}} = 3.22\ \text{mV}\,. \tag{2.32}$$

Comparing the value of the LSB in Eq. (2.32) to the standard offset voltage of CMOS operational amplifiers, which is in the range of $5 \sim 15\ \text{mV}$ [3], it can be seen that the design of a 10-bit A/D converter in flash architecture is still a challenge. One way to reduce the offset is to employ offset-reduction techniques, such as comparator auto-zeroing. Another way is to employ a flash A/D converter with lower string resolution and interpolate the remaining values. The *interpolation design technique* employed between the analog input signal and the comparator array, as illustrated in Fig. 2.12, can thus improve the situation [4]. It also reduces the input capacitance of the flash A/D converter since the input capacitance of the preamplifiers required in the interpolation stage is lower than that of comparator array and the total number of the preamplifiers is reduced by a factor 2. But the total number of comparators remains the same.

A sample-and-hold amplifier for sampling of the input signal is not a necessary component for the flash A/D conversion. But, since the CMOS high-speed comparator usually contains a differential amplifier at its input, the insertion of a sample-and-hold amplifier in front of the comparator array can help avoiding improper signal racing among the differential amplifiers of the parallel connected high-speed comparators, reduce the input impedance, and increase the analog bandwidth of the whole conversion system.

2.5.2 Two-Step Flash A/D Converter and Subranging A/D Converter

Although the flash A/D converter represents the fastest analog-to-digital converter, its die area and power dissipation increase exponentially with the resolution. This limits the use of this type converter for resolution exceeding 8~10-bit. Hence, architectures that exploit the advantages of the flash A/D converter and at the same time improve the resolution of the conversion system without expanding the die area and the power dissipation exponentially have been developed.

A straightforward way to achieve this is to separate the total resolution into a coarse part and a fine part. Each part employs its own flash A/D converter: both parts are resolved in sequence, and the resulting digital outputs are then combined to form a final output signal. Therefore, such kind of A/D converter is named *two-step flash A/D converter* or *half-flash A/D converter*. The recent research performed by Mr. van der Ploeg et al. [21] yielded a 12-bit two-step ADC fabricated in $0.25\ \mu m$ – CMOS technology working under $54\ \text{MSPS}$ with $295\ \text{mW}$ at $2.5\ \text{V}$. Another implementation realized by Mr.

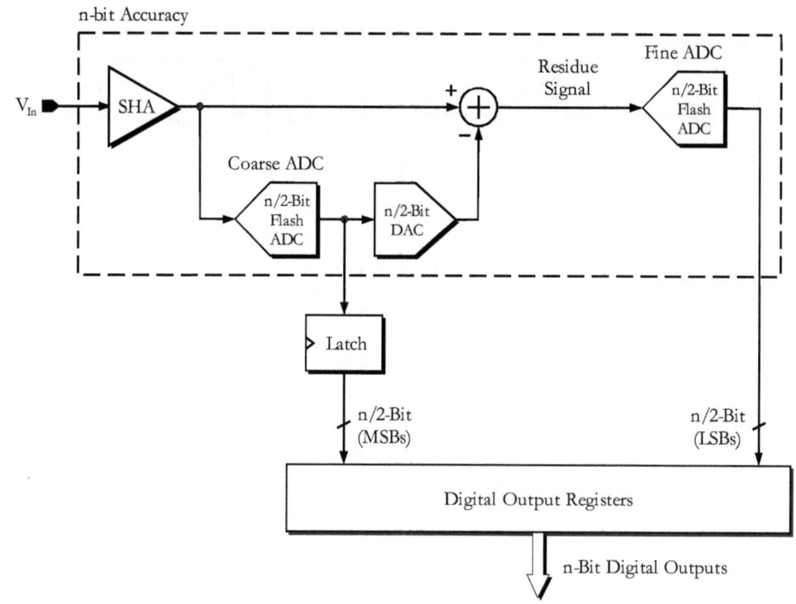

Fig. 2.13: Block diagram of a simple two-step flash A/D converter

Taft et al. [22] achieved an 8-bit two-step ADC working under 100 MHz and dissipating 84 mW at 2.2 V. This 8-bit two-step ADC has been fabricated in 0.35 m – CMOS process.

A block diagram of a two-step flash A/D converter exhibiting n-bit resolution is illustrated in Fig. 2.13. A sample-and-hold amplifier (SHA) at the front-end samples the analog input signal. This signal is held and fed into a coarse flash A/D converter, which resolves, for example, half of the total conversion resolution $n/2$. These $n/2$-bit output signals of the coarse flash A/D converter are fed into a $n/2$-bit digital-to-analog converter (DAC) and into a digital output latch in parallel. The D/A converter generates an analog signal, which represents only the magnitude of the coarse resolution while the digital output latch holds the results of the coarse resolution at first.

Subtracting the output signal of the D/A converter from the sampled signal yields a *residue signal*, which represents the magnitude of the fine resolution. Applying another flash A/D converter to the residue signal quantizes the last $n/2$ bits of the conversion system. Since this kind of A/D converter divides the total resolution into two parts, it is also known as *subranging A/D converter* in the literature.

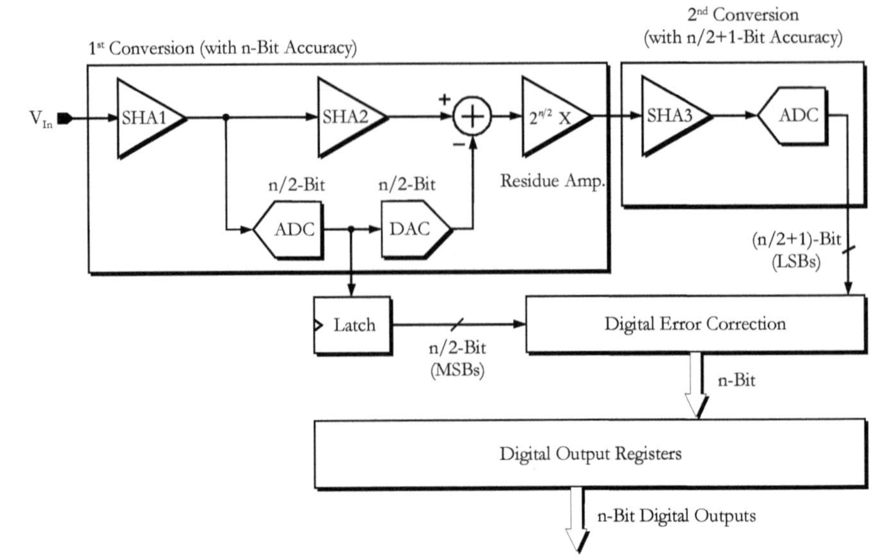

Fig. 2.14: Block diagram of a two-step flash A/D converter with error correction

For an *n*-bit converter as described above, the total number of high-speed comparators can be expressed as:

$$(2^{n/2}-1)\cdot 2 = 2^{(n/2+1)} - 2 \ll 2^n - 1, \quad \text{if } n > 2\,. \tag{2.33}$$

Although the conversion process of the two-step flash conversion is split into two parts, the accuracy of all comparators and other analog devices, as indicated in Fig. 2.13, must be *n*-bit to ensure that the total resolution of the conversion system remains unchanged. Thus, comparators designed with auto-zeroing design technique to reduce the offset in the comparator are necessary.

Compared to the conventional flash A/D converter, the two-step flash A/D converter needs a sample-and-hold amplifier and a digital-to-analog converter to generate the analog residue signal. The disadvantage of this architecture is that a high-speed sample-and-hold amplifier, which is one of the most difficult components of high-speed conversion, cannot be avoided. The sample-and-hold amplifier works as an "analog latch" at the front-end of the whole system and makes the analog subtraction with the output signal of the digital-to-analog converter possible.

The two-step A/D converter may exhibit problems, such as missing codes or nonlinearities, if the residue signal does not fit the input signal range of the second flash converter. To prevent this, methods of error correction have been developed.

Fig. 2.14 illustrates the block diagram of a two-step flash converter with *digital error correction* [4]. It is also known as *digitally corrected subranging A/D converter* (*DCS ADC*) in the literature. The first sample-and-hold amplifier holds the input signal for the coarse quantizer and the analog subtraction while the second sample-and-hold amplifier just compensates for the latency of both inputs of the analog subtraction. The third sample-and-hold amplifier provides "deglitching" of the residue signal. But it is not always necessary if the *residue amplifier* can drive the following ADC by itself without problems.

Due to the $2^{n/2}$-times amplification of the residue amplifier, the accuracy of analog signal processing in the 2nd conversion stage has to reach only $n/2+1$-bit accuracy, instead of n-bit accuracy, as indicated in Fig. 2.14 [23]. The accuracy in the first conversion stage still has to reach n-bit accuracy. Moreover, the resolution of the D/A converter applied in the first conversion stage must be better than the A/D converter used in the same stage to reduce the errors caused by the D/A converter. Generally speaking, if a D/A converter is connected in series with an A/D converter, the resolution of the D/A converter must be 5~8 times higher than the A/D converter itself. This equals to a resolution increase of about 2~3 bit. Thus, an n-bit A/D converter designed in two-step flash architecture needs a D/A converter of at least $(n+2)$-bit accuracy. Since the D/A converter is easier to design than the A/D converter, the resolution of the D/A converter shall not be the obstacle of the two-step flash architecture.

The second conversion stage provides one bit at its output more than the architecture without digital error correction. If no errors occur in the first conversion stage, the *most significant bit* (*MSB*) of the second conversion stays always "0". Apart from this, the MSB shows "1" and this information is sent to the following stage to correct the LSB from the first conversion.

This kind of digital error correction can correct errors that occur in the first $n/2$-bit converter. Errors occurring in the second conversion stage, in the D/A converter, or in the residue amplifier, cannot be corrected and are translated into the final outputs directly.

An inherent feature of the two-step flash A/D converter is that the digital output signals of the two-step flash A/D converter have latency of one clock cycle. Therefore, a digital latch must be inserted between the first conversion

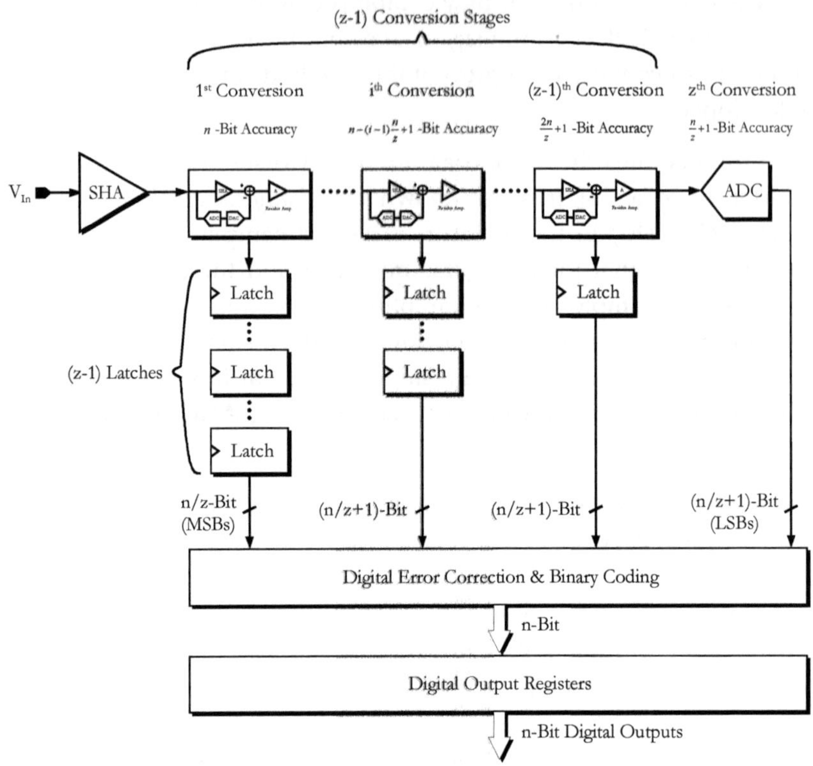

Fig. 2.15: Block diagram of a pipeline A/D converter

output and the digital output register (Fig. 2.13) to synchronize the both conversion stages. Since this latency is constant and can be clearly defined, it is acceptable for most of the applications.

Based on this principle of the two-step flash A/D converter, the architecture of *multi-step flash A/D converter* has been developed. This will be treated in the following section.

2.5.3 Multi-Step Flash A/D Converter and Pipeline A/D Converter

The *pipeline A/D converter* is a succession version of the two-step flash A/D converter (subranging A/D converter). In this kind of conversion architecture, the number of conversion stages has been extended in order to obtain a higher throughput. The pipeline ADC is one of the most popular architec-

Table 2.3: Component count for a pipeline A/D converter

Compo-nent	SHA	n/z -bit ADC	$n/z+2$ -bit DAC	$n/z+1$ -bit ADC	$n/z+3$ -bit DAC	Residue ampli-fier	Latch
Number	z	1	1	$z-1$	$z-2$	$z-1$	$\frac{z^2-z}{2}$

tures of high-speed A/D conversion. Recent research results can be found in [24-38]. An outstanding result is represented by a 14-bit pipeline ADC operating at powered by 75 MSPS while dissipating 340 mW at 3 V [39]. This pipeline ADC has been fabricated in $0.35\ \mu m$ – CMOS technology.

A block diagram of a typical pipeline A/D converter is depicted in Fig. 2.15. Essentially, the architecture of the pipeline A/D converter is quite the same as the architecture of the two-step flash A/D converter. The main difference is that there are several conversion stages between the first SHA and the last conversion stage. The number of the digital delay components has also been increased to cope with the increased number of conversion stages. Hence, the total latency of the conversion system has substantially increased.

The conversion stages of the pipeline A/D converter can be designed in such a way that they exhibit different accuracies as indicated in Fig. 2.15. For an n-bit pipeline A/D converter, the SHA at the analog front-end and the 1st conversion stage have to exhibit accuracy of n-bit. Let us assume that the system resolution n can be divided by the number of total conversion stages z and there is one bit overlap between each two neighboring conversion stages, as it is shown in Fig. 2.15. The 2nd conversion stage has only to maintain $n-n/z+1$-bit accuracy. The accuracy requirement for the ith conversion stage can be stated as:

$$n-(i-1)\cdot\frac{n}{z}+1, \quad \text{if } \frac{n}{z}\in\mathbb{N}\,, \tag{2.34}$$

where $\mathbb{N}$ represents set of the nature numbers.

To simplify further discussion, however, we shall assume that all conversion stages have identical n-bit accuracy. The total number of high-speed comparators can thus be written as:

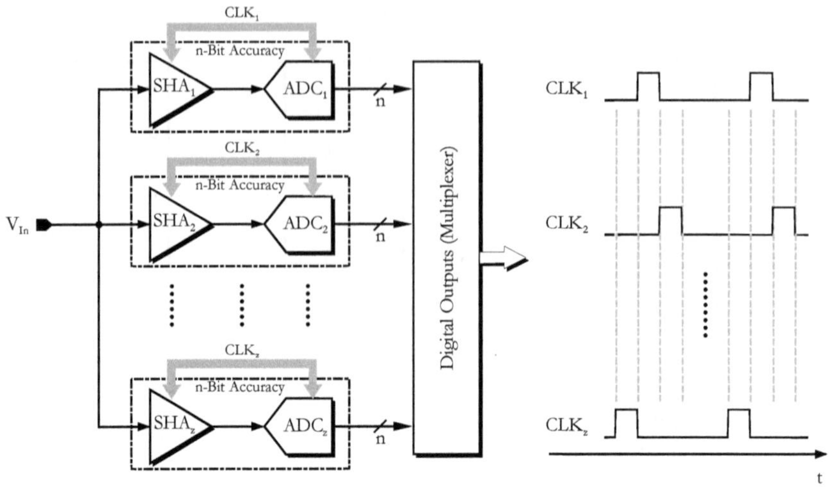

Fig. 2.16: Block diagram and clock timing of a time interleaved A/D converter

$$(2^{n/z}-1)+(z-1)\cdot(2^{n/z+1}-1),\ \text{ if } \frac{n}{z}\in\mathbb{N}\,. \tag{2.35}$$

The total number of latches is also a function of the parameter z and can be expressed as:

$$\sum_{i=1}^{z-1} i = \frac{1+(z-1)}{2}\cdot(z-1) = \frac{z^2-z}{2}\,. \tag{2.36}$$

Table 2.3 summarizes details of the component count for a pipeline A/D converter. However, it is difficult to express exactly the difference between the architecture of the flash A/D converter and the pipeline A/D converter without addressing other system parameters, such as die area and power dissipation, for each of the components. But in terms of the component count the pipeline A/D converter is clearly superior to the flash A/D converter. This is owing to the fact that the component count for the flash A/D converter rises exponentially with the resolution while the component count rises linearly for the pipeline A/D converter.

Similarly as in the case of the two-step flash A/D converter, the digital error correction can only rectify processing errors occurring in the interstage conversion. Internal A/D conversion gain, offset, and linearity errors can be corrected, if the amplified residue signal is inside the range of the next conversion stage. Errors occurring in the last conversion stage cannot be corrected and will be translated into the final digital outputs.

The conversion rate of the single conversion stage determines the conversion rate f_{AD} of the whole pipeline converter. This is typical of any pipeline processing. Since the latency of each conversion stage may differ, the maximum of the system conversion rate f_{AD} has to be set to the reciprocal value of the largest single-stage latency to compensate the different latencies. Introducing different clocks may reduce the conversion time for each conversion stage, thus increasing the conversion rate f_{AD} and, subsequently, the throughput.

2.5.4 Time Interleaved A/D Converter

A further possibility to achieve a high conversion rate f_{AD} is to use the architecture of the *time interleaved A/D converter* [40]. The block diagram of a time interleaved A/D converter and its clocking schedule are depicted in Fig. 2.16.

For a time interleaved A/D converter consisting of z twigs, each twig has its own SHA and ADC, which can be based on flash architecture. The analog input signal is applied in parallel to all of the SHAs at the same time, but the switching clocks of the SHAs and the following ADCs are set to be "active" one after the other, as shown in Fig. 2.16. The digital ADC outputs are multiplexed and combined in such a way that they form the right output signal from output signals of the conversion twigs. From the viewpoint of data processing, this conversion architecture represents intrinsically kind of parallel processing because the conversion is carried out consecutively by different circuits. Since the conversion twig operates only when their clock signal is "active", the conversion twig does not operate fully in parallel.

An n-bit time interleaved A/D converter of z conversion twigs, which has conversion rate of f_{Clk}, needs z SHAs and z flash ADCs of n-bit accuracy and can reach conversion rate f_{AD} of $z \cdot f_{Clk}$. Consequently, due to the multiple of the flash A/D converters, the time interleaved A/D converter offers no advantages concerning the issue of die area.

Another disadvantage of the time interleaved A/D converter is the gain and offset errors caused by the mismatch among different conversion twigs. Besides, the timing problem among the twigs can be getting worse when the conversion rate of each twig f_{Clk} increases [40].

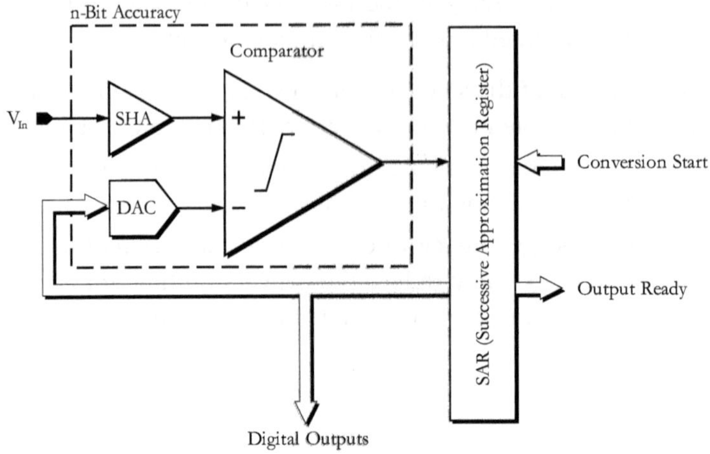

Fig. 2.17: Block diagram of an SAR A/D converter

Based on the discussion above, it can be concluded that this kind of conversion architecture is only appropriate for low-resolution high-speed applications.

2.5.5 Successive Approximation Register A/D Converter

The architecture of *successive approximation register A/D converter* (*SAR ADC*) has been used for decades because of its simplicity and small die area [4]. It is based on a "weighting" concept. The block diagram of a simplified successive approximation register A/D converter is illustrated in Fig. 2.17.

At the beginning of the conversion process, the *MSB* of the DAC input signal is set to "1" and the rest bits of the DAC input signal are set to "0", in order to place the DAC output signal to the middle voltage of the conversion range. At the same time, the analog input signal is sampled using a SHA front-end component. During the first clock cycle, the SHA output signal and the DAC output signal are compared by a comparator. The comparison result sets the *MSB* to either logical "0" or "1" depending on which voltage, i.e. DAC or SHA output, is higher and this *MSB* is stored in the register.

At the next clock cycle, the first comparison result (i.e. *MSB*) is read out from the register and the next most significant bit MSB_{-1} is now set to "1" with the rest bits of the DAC input signal staying at "0". These values are now fed into the DAC input signal to set the reference signal level for the 2nd comparison. The result of the 2nd comparison sets MSB_{-1} in the register.

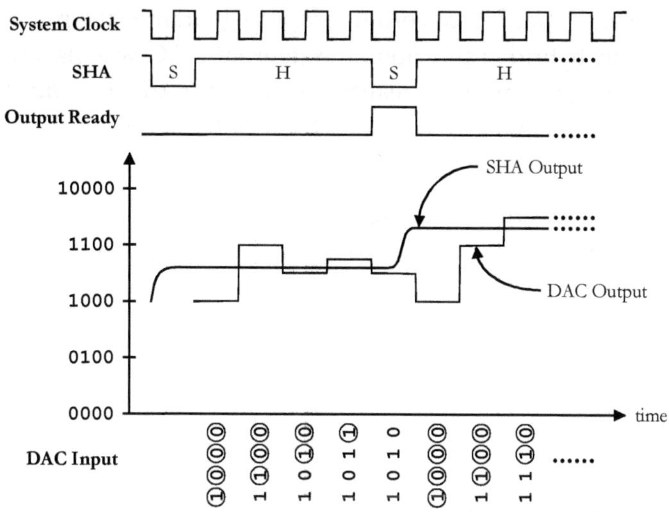

Fig. 2.18: Clock schedule and approximation procedure of a 4-bit SAR converter

This process is repeated until all remaining bits have been determined and stored in the register. When the iterations have been completed, the control bit "Output Ready" is set to logical "1" and the DAC input signal shows the final conversion result at this moment. Since the content of the register approximates successively the final conversion result, this kind of conversion architecture is called *successive approximation register conversion*.

Fig. 2.18 shows the clock timing and the approximation procedure of a 4-bit SAR converter. The encircled digits in the DAC input signal indicate bits that are to be determined using the iterative process described above. The bits that are not encircled represent the comparison results that have been already determined.

The greatest advantage of this architecture is that the conversion system requires a minimum of hardware because there is only a single high-speed comparator. Thus, the die area is very small. Besides, the system design is quite simple. Only the front-end SHA represents a problem if a high-speed operation is required. Otherwise, the other circuits, like the register and the DAC, are not very critical. For an n-bit SAR ADC, the devices, including the SHA, the comparator, and the DAC, have to reach at least the system resolution n-bit, as indicated in Fig. 2.17. The performance of the internal DAC affects both the accuracy and the conversion rate f_{AD} of the SAR A/D converter. Thus, the internal DAC should have $(n+2)$-bit accuracy and keep the last 2 least significant bits at zero in order to guarantee the resolution of the ADC itself. In hybrid realizations the internal DAC was usually based on weighted voltages: this was realized using *laser trimmed thin film resistor strings* employed as voltage dividers. Most of the recent designs use *CMOS switched capacitor charge redistribution DACs* instead [4].

For an SAR converter featuring *n*-bit resolution, a full conversion process needs $(n+1)$ system clock cycles. Hence the conversion rate f_{AD} is inherently lower than that of high-speed A/D converter architectures described above. The conversion rate f_{AD} can be expressed as:

$$f_{AD} = \frac{f_{Clk}}{(n+1)} . \tag{2.37}$$

For monolithic integration the maximum clock frequency f_{Clk} is usually determined by the technology used for the realization. For a given conversion rate f_{AD}, the required clock frequency can be rewritten as:

$$f_{Clk} \equiv (n+1) \cdot f_{AD} , \tag{2.38}$$

and

$$(n+1) \uparrow \Leftrightarrow f_{AD} \downarrow \quad \text{or} \quad (n+1) \downarrow \Leftrightarrow f_{AD} \uparrow . \tag{2.39}$$

Eq. (2.39) shows that the SAR A/D converter is suitable either for applications requiring high resolution at lower conversion rates f_{AD} or applications with low resolution at higher conversion rates f_{AD}. Thus this type of A/D converter is only useful when a trade-off between the resolution and the conversion rate f_{AD} can be found.

2.5.6 Folding and Interpolating A/D Converter

The *folding and interpolating A/D converter* has been under development since late 1970s and the original concept was developed for integration in bipolar technologies [41-46]. Recent research has focused on the development of CMOS folding and interpolating A/D converters and the results can be found in [32, 47-70]. For low-resolution applications, 6-bit folding and interpolating A/D converters integrated in CMOS have reached the conversion rate of 400 MSPS [51, 52], while an 8-bit implementation has reached the

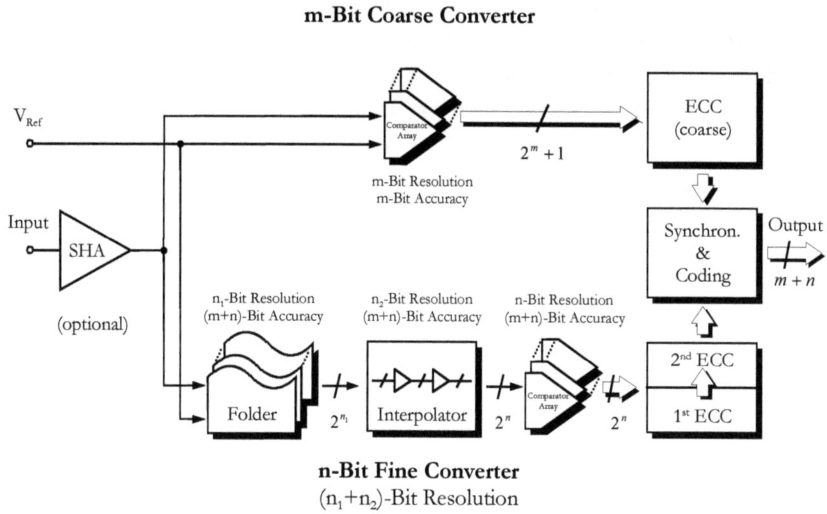

Fig. 2.19: Block diagram of an (*m*+*n*)-bit folding and interpolating A/D converter

conversion rate of 125 MSPS [61]. Moreover, the folding and interpolating A/D converters exhibiting 12-bit and 13-bit resolutions have been realized with the conversion rates up to 50 MSPS [32] and 40 MSPS [65, 66], respectively.

Most of the implementations were employed *voltage-mode interpolation*. Few of them were based on *current-mode interpolation* [7, 47, 51, 52, 67, 70].

A simplified block diagram of a folding and interpolating A/D converter contains an optional sample-and-hold amplifier, an *m*-bit *coarse converter*, an *n*-bit *fine converter*, and a *digital synchronization stage*, as illustrated in Fig. 2.19. Note that the sample-and-hold amplifier, which is one of the most critical devices in high-speed conversion, is not necessarily needed and represents only an option [10].

If employed at the front-end of the folding and interpolating A/D converter (Fig. 2.19), the optional sample-and-hold amplifier works as an *analog front-end*. If the input frequency f_{In}, compared to the conversion rate f_{AD}, is quite high, use of the sample-and-hold amplifier at the front-end can compensate for different latencies caused by the coarse and fine converters and share the synchronization task with the digital synchronization stage. Of course, this causes an increase of the die area and power dissipation.

The folding and interpolating A/D converter contains two separate converters, namely, the coarse and the fine converter. The input signal is fed into both converters in parallel. For a converter exhibiting total $m+n$-bit resolution, the *m*-bit coarse converter contributes the bits of the higher *m*-bit resolution, beginning from the *MSB* of the whole A/D converter. The *n*-bit fine converter generates the remaining lower *n*-bits, which ends with the *LSB* of the whole A/D converter.

An intuitive explanation, advanced by Mr. Bult et al. [71, 72], explains the relationship between the coarse and the fine converter as the relationship between the hour hand and the minute hand of a clock. In case that the analog clock has only minute hand, the clock must be segmented into 1440 parts to have the accuracy of a minute. On the other hand, the hour hand of a clock divides a day into 24 intervals and each interval is again divided into 60 parts by the minute hand. Thus, only 84 "decisions" have to be made.

The coarse converter can be realized as a simple flash architecture. It divides the input signal range into several intervals. On the other hand, the fine converter is composed of several stages, including the *folding stage*, the *interpolation stage*, the *comparator stage* (or *comparator array*), and the *digital error correction and coding stage* (ECC). The output signals of both converters are thermometer codes, which can be processed by the following digital circuits. This kind of resolution combination will be noted as m/n-architecture through this book [48].

Furthermore, since both the folding stage and the interpolation stage contribute to the conversion resolution, the notation of the different architectures of the folding and interpolating A/D converter has to reflect this. Thus for an m/n-architecture, as depicted in Fig. 2.19, we can designate the architecture as $m/n_1/n_2$-architecture instead to indicate the *folding resolution* n_1 and the *interpolation resolution* n_2.

Another advantage of the folding and interpolation A/D converter is that this architecture reduces the total number of high-speed comparators needed, especially for the resolutions exceeding 8-bit. A folding and interpolating A/D converter of $m+n$-bit resolution needs $(2^m-1)+2^n$ comparators, instead of $(2^{m+n}-1)$ for a flash A/D converter.

For a folding and interpolating A/D converter exhibiting a resolution of $m+n$-bit, as indicated in Fig. 2.19, analog parts of the fine converter, including the folding stage, the interpolation stage, and the comparator stage, have to reach the whole system accuracy $m+n$-bit despite of their different contributions on the system resolution. But, on the contrary, the accuracy of comparators in the *m*-bit coarse converter have to exhibit only *m*-bit accuracy when properly synchronized. Due to the accuracy difference, comparators in the coarse converter will be called *coarse comparators* and, on the contrary, comparators in the fine converter will be called *fine comparators*. Detailed description about how the synchronization stage works will be given in the following chapter.

Thus, not only the total number of high-speed comparators of the folding and interpolating A/D converter is reduced, but also the accuracy of the coarse comparators is lower than those of other converter architectures. Moreover, since the input signals of comparators of the fine converter are differential signals, comparators of the fine converter are less sensitive to noise than comparators, for example, of the flash A/D converter. This also leads to simplification of design of the comparators for the folding and interpolating A/D converter.

The comparators for the flash A/D converter are *"level comparators"*, which compare the input signal with a predefined reference voltage levels. Such comparators must exhibit very low noise and offset voltages and thus noise and offset reduction techniques must be employed for their design, e.g. *auto-zeroing*. To the contrary, the comparators for the fine converter compare differential input signals, which are generated by the folding stage in the front of the fine converter, not the input signal with an absolute reference value. This kind of comparison is also called *"zero-crossing detection"* in the literature [7, 10, 16, 40, 47, 48, 51-53, 70-73]. Details of zero-crossing detection will be treated in Chapter 3.

However, since the coarse and fine comparators are not the only components of the folding and interpolating A/D converter, the die area and the power dissipation of the other components, especially the components of the fine converter such as the folding stage and the interpolation stage, must be also taken into account.

At the end of the signal path (Fig. 2.19), the results of the two converters are synchronized and combined together in a digital synchronization stage. The output signals of the converters can be coded into suitable binary code, for example the *straightforward binary code* or the *two's complement code*, and the output of the whole converter then exhibits $m+n$-bit resolution.

For a better understanding of how the folding and interpolating A/D converter works, let us now consider an 8-bit A/D converter as an example. An 8-bit A/D converter based on the flash architecture requires 255 high-speed comparators to resolve all of its representable states. For the purpose of dis-

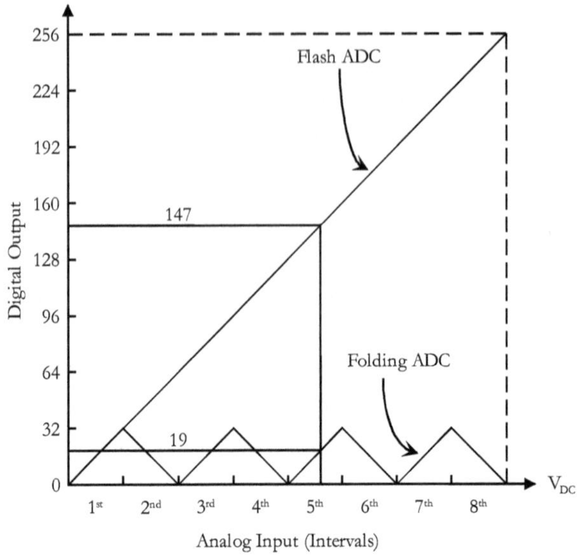

Fig. 2.20: Transfer functions between flash ADC and folding ADC

cussion, the staircase transfer function of this converter is depicted in Fig. 2.20 as a straight line, without loss of correctness.

Contrary to the 8-bit flash operation, the folding A/D converter preprocesses the input signal by dividing its range into, for example, 8 intervals if the coarse converter has 3-bit resolution. Hence, only 7 comparators in the coarse converter are necessary for indication in which interval the input signal is to be found, while the relative position inside the interval is determined by the 32 fine comparators. The total number of high-speed comparators is then 39, instead of 255. The state **147**, for example, can be indicated as **4**·32+**19**. This kind of design reduces substantially both the power dissipation and the die area of high-speed comparators for the high-resolution A/D converter.

A drawback of the folding and interpolating A/D converter, which is mentioned very often in the literature [47, 70, 74, 75], is the *internal frequency* generated by the folding stage. This problem can be better understood when considering the same 8-bit A/D converter in 3/5-architecture described above.

Providing that the input signal of the folding stage is a saw-tooth signal as depicted in Fig. 2.21 and it has a period T, the output signal of the folding stage shows the folded input signal, the period of which is only $T/8$. Hence, the output frequency of the folding stage is 8 times higher than its input

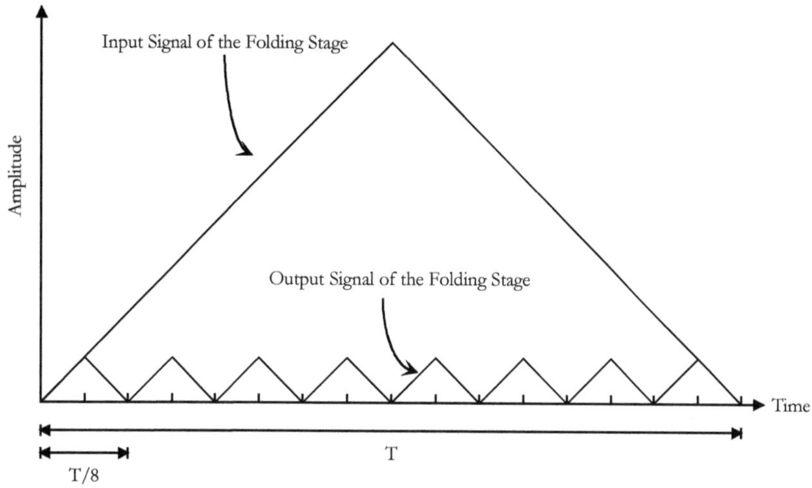

Fig. 2.21: Creation of the internal frequency by the folding stage

frequency. The increased frequency occurs after the folding stage; therefore, it is named "internal" frequency. Further, if the input signal is a sine-wave signal, the internal frequency reaches its maximum at the mid point of the sine-wave signal and it is by a factor of $\pi/2$ larger than that of a saw-tooth signal with the same frequency [70].

Using the SHA circuit, the input signal for the coarse and fine converters can be sampled and then "frozen", i.e. held at a fixed level. The effect of the multiplied internal frequency is then reduced and the sampled input signal works combination of step functions. The design problem is no more the increasing internal frequency, but fast step response and low latency of the signal flow in the converter [48]. More discussion about those topics and how the folding stage generates the folded input signal can be found in Chapter 3.3.

2.6 Problems of Embedded System with A/D Converter

The today's microelectronic industry represents an industry with extremely fast development cycles. The leading fabrication technology for microelectronic circuit is the CMOS technology. The state-of-the-art gate length of a CMOS transistor has shrunk already down to $0.11 \sim 0.13\ \mu m$ for digital circuits. The steady reduction of the device sizes enables fabrication of micro-

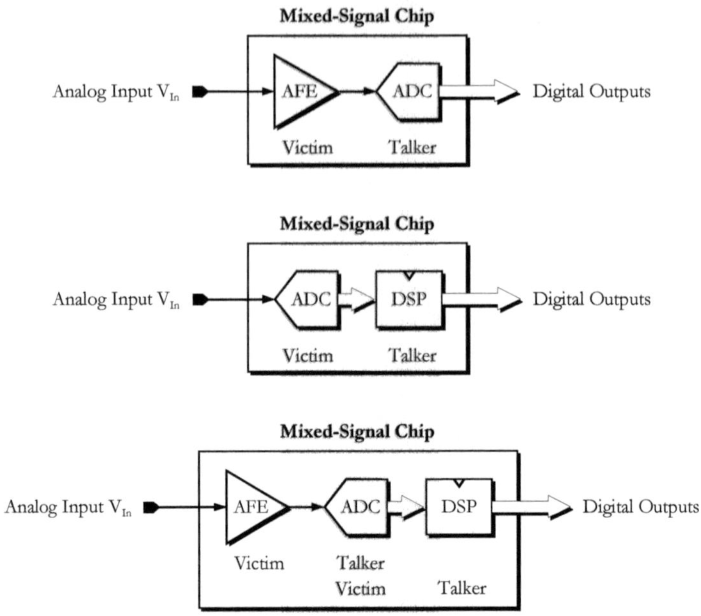

Fig. 2.22: Functions of the A/D converter in a mixed-signal chip

electronic circuits that are either getting smaller and smaller for the same functionality (i.e. chips containing a given transistor count are smaller and thus cheaper) or have growing functionality (due to higher device count) for the same chip area. In the latter case this means that it is possible to realized entire microelectronic systems on a single chip (system-on-a-chip). The analog-to-digital and digital-to-analog converters represent very important components for systems processing mixed-signals, i.e. analog and digital.

Considering different components of a system-on-a-chip processing mixed-signals, the A/D converter usually plays an essential role in conversion of analog signals into digital ones. Nevertheless, the presence of analog and digital signals on the same chip necessarily raises the question of crosstalk. It is thus a mandatory issue to be addressed during the chip design. We can distinguish three situations involving an on-chip A/D converter on a mixed-signal chip. As showing in the Fig. 2.22, the first situation is that the A/D converter is combined only with purely analog circuits forming its analog front-end (AFE). For this situation, the converter is the "talker" of the crosstalk and the AFE is the "victim". In case that the output of the A/D converter is connected to a digital signal processor (DSP), the converter plays the role of a "victim" and the DSP itself is the "talker". If there are both AFE

Table 2.4: Summary of architectural characteristics of high-speed A/D converters

Architectures of A/D converters	Characteristics
Flash A/D converter	+ Fast architecture + High-speed SHA unnecessary – Complex high-speed comparator with auto-zeroing technique – Large number of high-speed comparators – Sensitive to the noise in the circuit noise and device mismatch – Large input impedance – Large
Two-step flash A/D converter (subranging A/D converter)	+ Fast architecture + Less high-speed comparators than flash ADC + High-speed, hgih-gain residue amplifier unnecessary + Low latency and less digital logic than multi-step flash ADC – High-speed SHA necessary – Complex high-speed comparator with aut
Multi-step flash A/D converter (pipeline A/D converter)	+ Fast architecture + Less high-speed comparators than flash and two-step ADCs + Less accuracy required in later stages + Capable of higher resolution – High-speed SHA necessary – Complex high-speed comparator with auto-zeroing technique
Time-interleaved A/D converter	+ Very fast architecture • Appropriate for low-resolution, high-speed applications – High-speed SHA necessary – Complex high-speed comparator with auto-zeroing technique – Large number of high-speed comparators – Large die size – More power dissipa
SAR A/D converter	+ Minimum of hardware + Low power dissipation + Only one high-speed comparator is necessary – Complex high-speed comparator with auto-zeroing technique – Slow conversion rate because of iteration process – High-speed SHA necessary
Folding and interpolating A/D converter	+ Fast architecture + High-speed SHA unnecessary + High-speed comparator with zero-crossing detection + Less high-speed comparators than flash ADC + Small die size + Less power dissipation – Synchronization between coarse and fine converters necess

and DSP on the single chip, the ADC is both "victim" and "talker" of the crosstalk.

Concerning the noise caused by crosstalk between the analog and digital circuits, the first case in Fig. 2.22 has much lower content of digital circuits than the other two cases. Therefore, the crosstalk effect in this type of embedded systems is smallest. The analog front end can be, for example, CMOS image sensor or other kinds of analog sensors.

2.7 Summary

In this chapter, concepts of analog-to-digital and digital-to-analog conversions were introduced. The sampling theorem which is so important for the A/D conversion was also discussed. Apart from quantization process, other effects have also impact on the performance of the A/D converters. Thus, device mismatch, crosstalk, and noise affect the static performance, like both, differential and integral nonlinearities, and the dynamic performance, such as SNR, THD, and SFDR.

The time-interleaved A/D converter yields the fastest conversion rate (see Table 2.4) because it uses several flash A/D converters in parallel. Consequently, A/D converters realized in this architecture are not suitable for the low-power applications. The flash A/D converter has similar problems. On the other hand, the successive approximation register A/D converter uses only minimum of hardware and, hence, it is a good candidate for the low-power applications. However, the conversion rate of this architecture is slowest due to its iteration conversion process.

The two-step or multi-step flash A/D converters belong to the group with fast conversion rates and requires lower count of high-speed comparators than the flash A/D converter. The two-step flash A/D converter can be implemented without high-speed, high-gain residue amplifiers (Fig. 2.13). The multi-step flash A/D converter requires fewer high-accuracy comparators than the two-step flash A/D converter, but high-speed, high-gain residue amplifiers cannot be avoided [76]. Both converters need a high-speed SHA, which represents a tough task in high-speed analog circuit design.

On the other hand, the high-speed SHA can be an optional component for the folding and interpolating A/D converter where it can help relax design requirements on the rest of the converter. Also, this type of converter requires only comparators performing zero-crossing detection to make their decision, unlike comparators serving as level comparators in other A/D converters. This leads to simplification of the design of the comparators for the folding and interpolating A/D converter.

Due to the advantages mentioned above, the folding and interpolating A/D converter has been chosen as a basis for development of a family of folding and interpolating A/D converters exhibiting high-speed and low-power, which are suitable for embedded applications.

CHAPTER 3

COMPONENTS FOR CMOS FOLDING AND INTERPOLATING A/D CONVERTERS

Epitome

In this chapter the individual components necessary for building folding and interpolating A/D converters will be described and analyzed in detail. The generic architecture of the folding and interpolating A/D converter is based on that of Fig. 2.19 in Chapter 2.5.6.

3.1 Subsystem: Sample-and-Hold Amplifier

3.1.1 Introduction

There are basically two sample-and-hold amplifier concepts, which are sometimes confused in practice. First is that of classical *sample-and-hold am-*

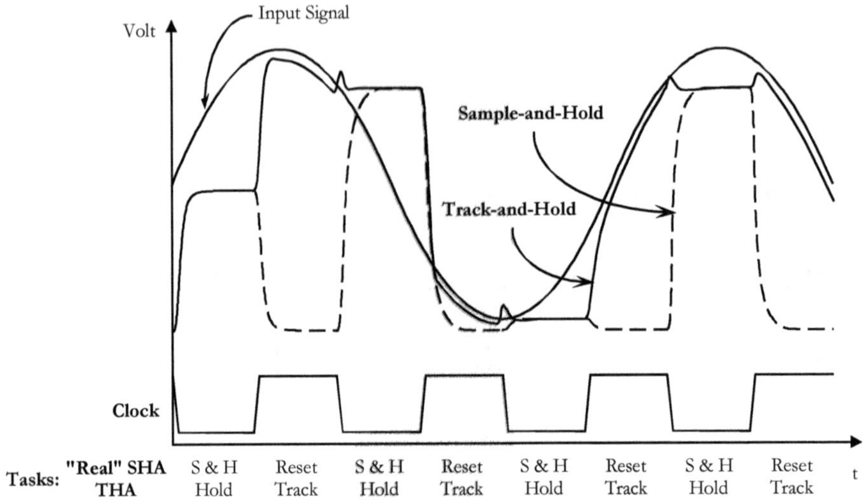

Fig. 3.1: Difference between "real" SHA and THA

plifier (*SHA*) which is the "real" sample-and-hold amplifier and then the *track-and-hold amplifier* (*THA*). There is a loose distinction between these two which can be briefly described in the following way: the "real" SHA indicates a circuit that has much longer *hold phase* than the *sample phase,* while the THA is a circuit with long *track phase* and a long hold phase [14]. The sample phase and the track phase are also called *acquisition phase* in the literature.

Generally speaking, the "real" SHA provides instantaneous sampling of the input signal (i.e. with infinitesimal sampling period) and the output is then held at this level for the rest of the sample period. The SHA is then reset during the reset phase. On the other hand, the output signal of the THA is forced to follow the input signal during the track phase and the last track signal is held during the following hold phase. Their difference is shown in Fig. 3.1. In this book, the concept of the "general" SHA will be used to identify all circuits that transform the input signal into a step-like signal, i.e. as a generic term. The "real" SHA will then be called "*narrow-defined SHA*". The concept of the THA will be used instead only for those circuits whose output signal coincides partially with the input signal, i.e. a more specified term [77]. A summary about these definitions is listed in Table 3.1.

There are some distinguishing points to identify the different performance of the narrow-defined SHA and the THA. At the beginning of the sample-and-hold phase of the narrow-defined SHA, the output voltage of the SHA must be charged from the ground potential to the input signal level, as

Table 3.1: Definitions of sampling methods

General SHA	Circuits, which transform the input signal to step-like signal.
Narrow-defined SHA	During the sample phase, the output signal is sampled and held.
THA	During the track phase, the output signal is forced to track the input signal.

shown in Fig. 3.1. This might cause a higher slew rate than that of the THA, since the output voltage of the THA tries to keep the same voltage level as the input signal during its track phase. Thus, our attention will focus on the THA in the following discussion.

The track-and-hold amplifier stage is typically the bottleneck for high-speed conversion. Basically, the THA has one or more capacitors that serve to capture and store the input signal. During the operation of the A/D converter, these capacitors are continually charged and discharged in order to reach the voltage level of the input signal. The time for charging and discharging these capacitors is called *acquisition time* t_{Acq}, as illustrated in Fig. 3.2. Besides this, the *settling time* t_{Set} in the hold phase is also an important criterion when evaluating an THA component. It describes the period during which the THA needs to stabilize its output signal at its final value. The settling time t_{Set} is usually defined from the beginning of the hold phase till the status when the difference between the voltage of the output signal and the target voltage is less than 0.1%. Other important definitions of the THA, such as *error band, pedestal,* and *droop rate,* are shown in Fig. 3.2 [77, 78].

The first consideration of inserting an THA circuit at the analog front-end is to capture and hold, i.e. to "freeze" the input signal for the following analog processing. As mentioned at the beginning of this chapter, the folding and interpolating A/D converter can work without an THA circuit at its front-end. But, since the folding stage in the fine converter contains several parallel-connected differential amplifiers at its input, varying signal magnitudes at their input sides can cause different operating speeds among the various differential amplifiers. Based on this, a *signal racing* arises in the folding stage, which might cause errors in the following stages. To avoid such situation, it is preferable to insert an THA circuit in front of the differential amplifiers to work as a "pacemaker" [48].

This signal flow in the fine converter includes the folding stage, the interpolation stage, and the comparator stage. The processes of interpolation and

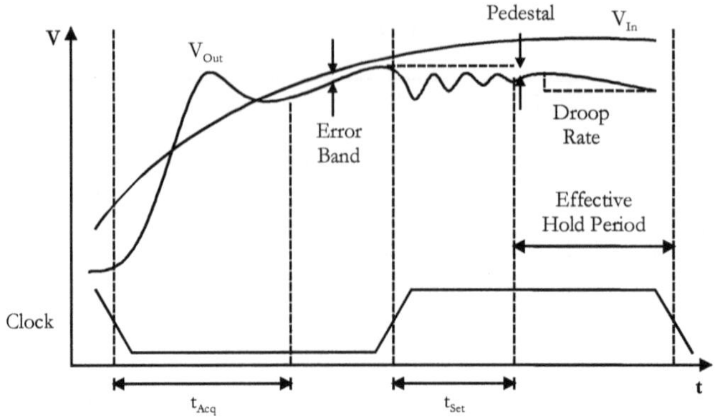

Fig. 3.2: Common definitions for track-and-hold amplifier

comparison are faster than that of the folding stage. Considering a 40 MSPS A/D converter with a standard sample-and-hold amplifier as its analog front-end, the hold period for the folding stage is theoretically max. 12.5 $n$sec, providing that the duty cycle of the clock signal is 50%. Nevertheless, the settling time t_{Set} of the THA circuit must be accounted for. At the end, the *effective hold period,* as defined in Fig. 3.2, will be less than 10 $n$sec, which may be insufficient to drive the next stages, if the power dissipation of the following stages is restricted.

Although a THA may have longer effective hold period than the narrow-defined SHA, it is usually still not enough. Thus, a more robust solution must be found. However, the high-speed SHA or THA circuit is not a trivial component for the analog circuit design and its power dissipation can be significant for high-speed applications. Although such circuit has a design history of over 20 years in CMOS, its design still represents a formidable task.

The following sections will discuss the different architectures of the THA circuit at first and then introduce the corresponding design techniques.

3.1.2 Principle of Double Sampling Design Technique

The latency of the signal flow in the fine converter is one of the limitations of the total conversion speed. If the clocks used in the THA circuit and in the comparators are coupled together, the signal propagation from the output of the THA stage to the output of the comparator stage of the fine converter must occur within a single clock cycle. Thus, the duration of the effective

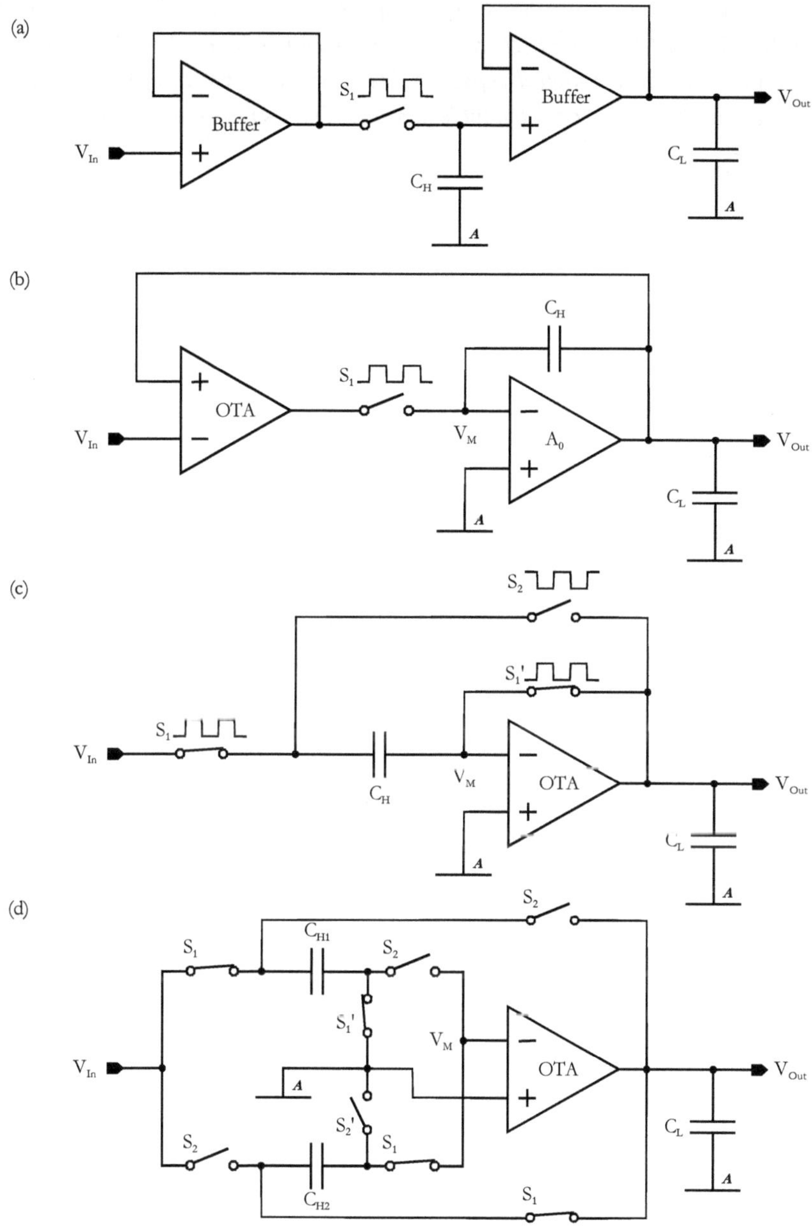

Fig. 3.3: Schematics of various architectures of THAs

track phase, which reflects the result of trade-off between the conversion rate and power dissipation, plays an important role.

Fig. 3.3 shows schematics of various architectures of THAs [77-79].

Fig. 3.3(a) is the *conventional open-loop architecture*. Fig. 3.3(b) shows the *conventional closed-loop architecture*. The next schematic, i.e. Fig. 3.3(c), depicts the architecture of a *switched-capacitor THA* and the last one, Fig. 3.3(d), relates to the architecture of the *double-sampled THA* suggested by [80], [81].

The open-loop architecture (Fig. 3.3(a) [77]) has a simple construction and its sampling speed is higher than that of other architecture because of its open-loop arrangement. As the switch S_1 is on, the circuit is in the track phase and the output signal V_{Out} tracks the input signal V_{In}. While the switch S_1 is turned off, the circuit is in the hold phase. The output signal V_{Out} corresponds to the stored input signal at the *hold capacitance* C_H. This architecture can be designated as a classical track-and-hold amplifier.

The sampling speed of this THA depends both on their acquisition time t_{Acq} and settling time t_{Set}. The acquisition time t_{Acq} is determined by the tracking speed and output impedance of the first buffer, the *on-resistance* R_{On} of the switch S_1, and the hold capacitance C_H. The settling time t_{Set} is mainly determined by performance of the second buffer.

Concerning the issue of the linearity, since both buffers are connected as a *unity-gain amplifier*, the *input signal range* and the *output signal swing* of both buffers play important roles. Trade-off between the sampling speed and the system linearity has to be made. Besides, the hold capacitance suffers from *charge injection* caused by the sampling switches. This also contributes to the nonlinearity of CMOS implementations. Furthermore, this architecture contains two unity-gain buffers, which is prohibitive for low-power designs.

In summary, the open-loop architecture includes no global feedback, thus, the circuit is stable. This topology can reach high-speed sampling, but suffers from *input-dependent pedestal error* for the CMOS implementations [77].

Fig. 3.3(b) is a schematic of the conventional closed-loop architecture [77]. As the switch S_1 is on, the circuit is in the track phase. The node V_M represents the *virtual ground* and the capacitance C_H works as a *frequency compensation capacitor* of a two-stage operational amplifier, where an OTA operates as the input stage. Note that both stages of the operational amplifier are configured as unity-gain amplifier during the track phase. If the amplification A_0 is large enough, the output voltage V_{Out} tracks the input voltage V_{In}. While the switch S_1 is turned off, the feedback circuit containing A_0 and C_H maintains the stored voltage at the output.

While the circuit in the track phase, both the output voltage of the *operational transconductance amplifier* (*OTA*) serving as the input stage and the

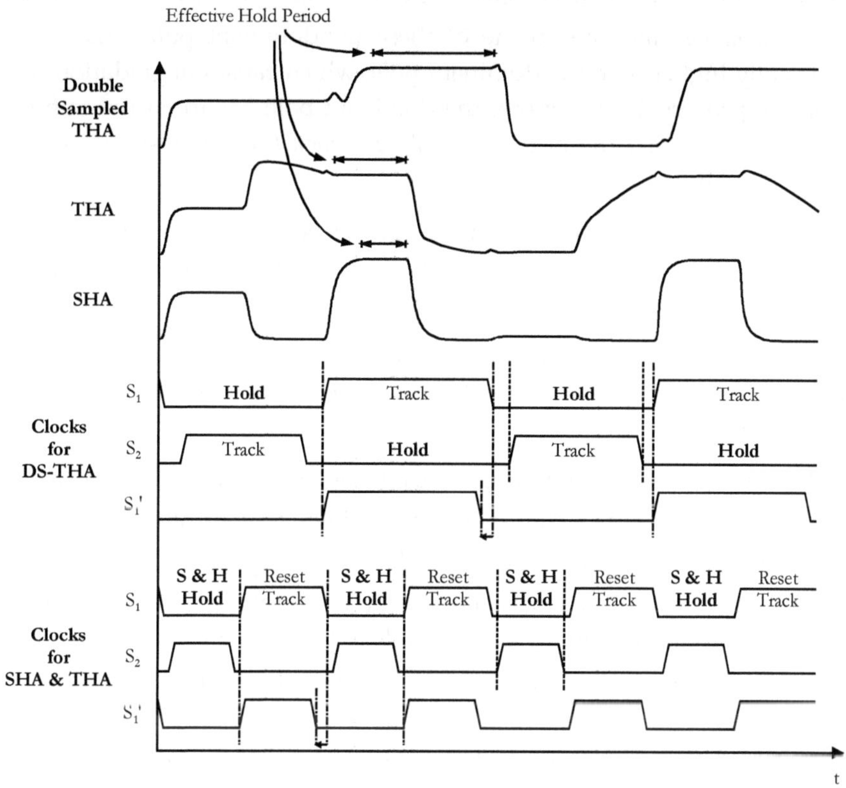

Fig. 3.4: Effective hold periods of double-sampled THA, THA, and SHA

virtual ground V_M, i.e. both the input and output signals of the switch S_1, are close to the ground potential. Hence, the charge injection caused by the switch S_1 is independent of the input signal V_{In}. The effect of charge injection acts like an offset of the sampled signal, i.e. like a pedestal error, and its nonlinearity contribution is insignificant. This offset can be even eliminated using a switch and a hold capacitance connected parallel between the noninverting input of A_0 and the ground. This adds a common-mode voltage to the noninverting input of A_0 to compensate the pedestal error caused by the charge injection of the switch S_1 [77].

Two main drawbacks of this architecture are the low speed and its potential instability. Since this architecture is like a two-stage op amp during its track phase, the *dominant pole* is determined solely by the OTA and the *Miller multiplication* of the hold capacitance C_H acting as a frequency compensation capacitor. But, both amplifiers, i.e. OTA and A_0, generally introduce several *nondominant poles*. Some of these nondominant poles may not be sufficiently higher than the dominant pole, which causes degradation of the sampling process. Furthermore, the dominant pole has to ensure sufficient *phase margin* so that the output signal can track the input signal with the required accuracy.

Although this architecture can reduce the pedestal error caused by the charge injection, its weakness are low speed and high power dissipation. This makes it unsuitable for our demands.

The switched-capacitor THA, illustrated in Fig. 3.3(c) [77, 78], uses only one amplifier in the circuit. This makes it a convenient circuit for low power applications.

This circuit has two operating states. In one state, while the switches S_1 and S_1' are closed and the switch S_2, which is controlled by the *non-overlapped clock signal,* stays open, one plate of the capacitor C_H is at "virtual ground" due to the feedback action of the operational transconductance amplifier and the voltage V_M corresponds to the virtual ground. The capacitor C_H is charged to the input signal V_{In}. The whole circuit is in the track phase. In the other state, the switches S_1 and S_1' are open and the switch S_2 is closed. In this case, the voltage stored at the capacitor C_H appears at the OTA output as the voltage V_{Out}. The circuit is in the hold phase.

The clock signal used to open the switch S_1' must act slightly earlier than that for the switch S_1 to prevent the charge injection, caused by the clock signal S_1, onto C_H. The corresponding clock scheme can be found in Fig. 3.4. In addition, as the switch S_1' switches between the output signal and the virtual ground V_M, its charge injection is independent of the input signal V_{In}.

The realization of this circuit is quite simple. If a single-stage operational transconductance amplifier is employed it can yield a linearity up to 13-bit [77, 82]. However, the effective hold period (Fig. 3.2) might be still insufficient for the following signal processing, especially, when low-power dissipation is required. The low-power operation increases latencies of the signal processing. Hence, longer effective hold period is preferred for low-power applications.

As noted above, in the track phase one plate of the capacitor C_H is at the virtual ground due to the feedback action of the OTA. Suppose that the capacitor C_H can be decoupled from the OTA and grounded, then the OTA can be used for other purposes. E.g., the OTA can be employed in a 2nd THA, which works in the hold phase. Such an arrangement is called *double-sampled THA (DS-THA)* in the literature. A basic architecture of the double-sampled THA suggested by [80], [81] is illustrated in Fig. 3.3(d). This topology can be divided into two parts. The upper part includes the capacitor C_{H1} and four switches, which are connected to the capacitor C_{H1}. The bottom part contains the capacitor C_{H2} and the other four switches. Each part operates similarly as the circuit depicted in Fig. 3.3(c), but at a double sampling rate with respect to the input signal V_{In}. The only difference is that, the capacitors C_{H1} and C_{H2} are "really" grounded and not "virtually" grounded during the sampling operation. But the clock signals applied to the switches S_1 and S_2 are non-overlapped. That is to say, while the upper part works in the track phase, the bottom part works in the hold phase, and vice versa. The OTA is only connected to that part, which is in the hold phase.

To maintain the same performance of the circuit illustrated in Fig. 3.3(c), the double-sampled THA can be clocked only at half of the clock frequency as used in Fig. 3.3. As a consequence of this, the effective hold period offered by the double-sampled THA is double that of the other architectures, which offers a benefit for the folding and interpolating A/D converter. A comparison of the different effective hold periods for double-sampled THA (DS-THA), THA, and classical SHA is shown in Fig. 3.4.

3.2 High-Speed Operational Transconductance Amplifier

3.2.1 Specification

The core component of the double-sampled THA is a high-speed *operational transconductance amplifier* (*OTA*). There are some criteria to evaluate the high-speed OTA. Among these criteria, the small-signal parameters, such as the *open loop gain* A_0, the *unity-gain frequency* f_{0dB}, and the *equivalent noise voltage*, and the large-signal parameters, such as the *slew rate SR*, the *output current* I_{Out}, the *settling time* t_{Set}, the *common-mode input range* $|V_{In}|$, and the *output voltage swing* $|V_{Out}|$, are of main importance.

Considering the situation of a simple THA, which operates at a sampling rate of f_{Clk} and exhibits a precision corresponding to n-bit accuracy. Main

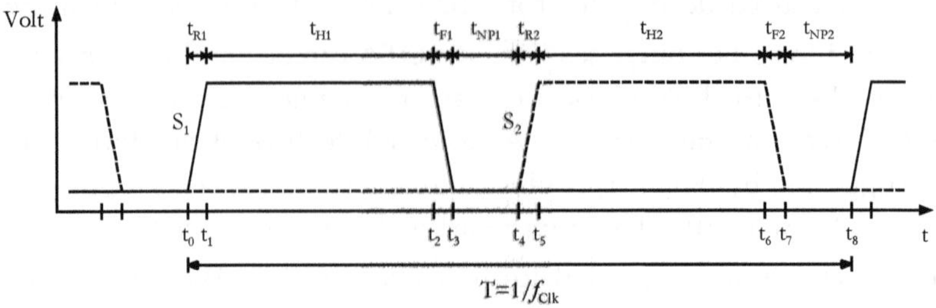

Fig. 3.5: Timing schedule of non-overlapped clock signals

parameters, such as the unity-gain frequency f_{0dB}, the slew rate *SR*, and the output current I_{Out} of the applied OTA, can be calculated as below.

The first issue is about the influence of the sampling rate on OTA parameters. Considering the clock signals, S_1 and S_2, as illustrated in Fig. 3.5, these two signals are non-overlapped clock signals. Since the sampling rate is f_{Clk}, the maximal useful duration *T* is $1/f_{Clk}$ as noted in Fig. 3.5. Furthermore, assuming that both the non-overlapped clock signals, S_1 and S_2, have also the following relationship:

$$t_{R1} = t_{R2} = t_{F1} = t_{F2} \text{ ,} \tag{3.1}$$

$$t_{NP1} = t_{NP2} \text{ ,} \tag{3.2}$$

$$t_{H1} = t_{H2} \text{ ,} \tag{3.3}$$

and

$$\begin{aligned} T &= \frac{1}{f_{Clk}} \\ &= t_{R1} + t_{H1} + t_{F1} + t_{NP1} + t_{R2} + t_{H2} + t_{F2} + t_{NP2} \\ &= 4t_{R1} + 2t_{H1} + 2t_{NP1} \text{ ,} \end{aligned} \tag{3.4}$$

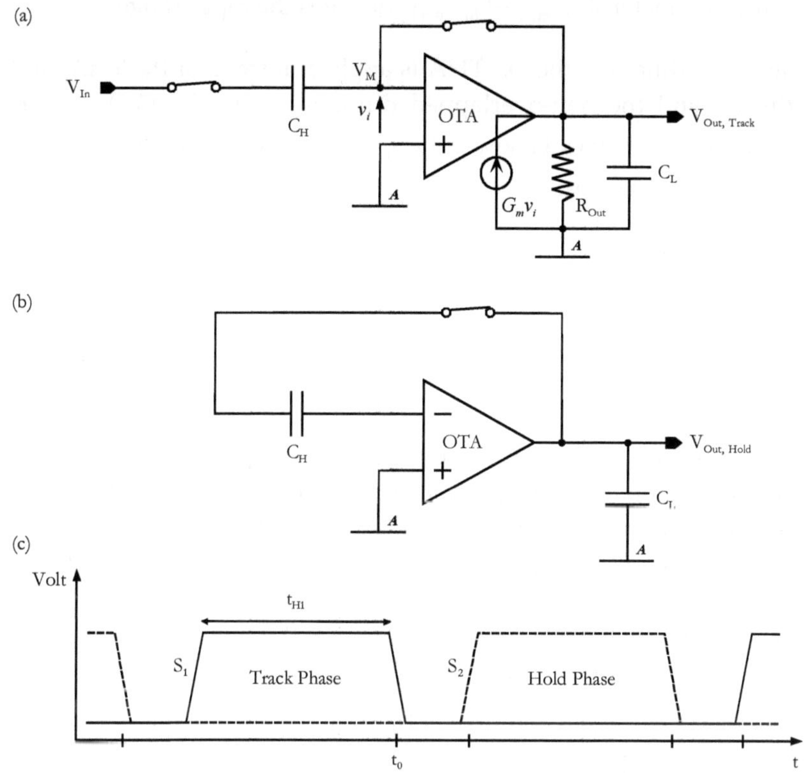

Fig. 3.6: Operating modes and time schedule of a THA

where t_R, t_F, t_{NP}, and t_H represent the *rise time*, the *fall time*, the *non-overlapped duration* and the *pulse width* of the clock signals, respectively. Thus, the pulse width t_{H1} can be described as:

$$t_{H1} = \frac{1}{2f_{Clk}} - 2t_{R1} - t_{NP1} \ . \tag{3.5}$$

Secondly, the impact of the designated accuracy is to be taken into account [83]. According to Eq. (2.16), the maximum quantization error ε is equal to half of an LSB. For an accuracy of *n*-bit, the maximum error V_{Error_max} can be described as:

$$\left|V_{Error\_max}\right| = \frac{1}{2} LSB = \frac{1}{2} \cdot \frac{\left|V_{In\_max} - V_{In\_min}\right|}{2^n} = \frac{\left|V_{In\_max} - V_{In\_min}\right|}{2^{n+1}} , \tag{3.6}$$

where the parameter $\left|V_{In\_max} - V_{In\_min}\right|$ indicates the input range.

Now, providing that the DS-THA is applied, instead of the THA, and the rise time t_R and the non-overlapped duration t_{NP} in Eq. (3.5) remain the same, the sampling clock is now only $f_{Clk}/2$. The pulse width t_H of the DS-THA can be described as:

$$t_H = \frac{1}{f_{Clk}} - 2t_R - t_{NP}\ , \tag{3.7}$$

where the first term is double than that of the pulse width t_H of the THA (see Eq. (3.5)).

The output load of the OTA in CMOS technology is usually only capacitive. The operational transconductance amplifier itself can be thus modeled as a voltage-controlled current source driving a load. We now model the THA of Fig. 3.3(c) for two phases separately while keeping in mind that this model applies to the circuit in Fig. 3.3(d), too. The circuit configuration during the track phase looks as shown in Fig. 3.6(a), where G_m is the transconductance and R_{Out} is the output resistance of the OTA, respectively. During the hold phase we obtain another configuration shown in Fig. 3.6 (b).

The output signal during the hold period can be determined as:

$$v_{Out,\,Hold}(t) = \frac{G_m R_{Out}}{G_m R_{Out}+1} \cdot V_{In}(t_0) \cdot \left[1 - e^{-\frac{(G_m R_{Out}+1)}{C_L R_{Out}}t}\right]. \tag{3.8}$$

$V_{In}(t_0)$ is the signal at time t_0 (see Fig. 3.6 (c)), i.e. the time instant when the switch S_1 is thrown off: this corresponds to the falling edge of the clock signal active during the track phase.

The output voltage during the track phase is

$$v_{Out,\,Track}(t) = \frac{C_H}{C_L + C_H} \cdot \frac{1}{G_m R_{Out}} \cdot e^{-\frac{(G_m R_{Out}+1)}{(C_L + C_H) R_{Out}}t}. \tag{3.9}$$

Note that the offset of the OTA and its low frequency $1/f$-noise are reduced for $G_m R_{Out} \gg 1$ due to cancellation effects. This can be shown considering an equivalent input offset voltage V_{OS} of the OTA shown above. The output voltage during the hold phase is then

$$v_{Out,\,Hold}(t) = \left[\frac{G_m R_{Out} \cdot V_{In}(t_0)}{G_m R_{Out}+1} + \frac{V_{OS}}{G_m R_{Out}+1}\right] \cdot \left[1 - e^{-\frac{(G_m R_{Out}+1)}{C_L R_{Out}} t}\right]. \quad (3.10)$$

The high frequency noise, however, is uncorrelated and cannot be cancelled. This will be considered in the following. But first we have to determine the noise bandwidth.

During the hold phase the 3dB-bandwidth is $f_{3dB,\,Hold} \cong G_m/2\pi C_L$ for $G_m R_{Out} \gg 1$, while during the track phase it is valid $f_{3dB,\,Track} \cong G_m/2\pi(C_H + C_L)$, again for $G_m R_{Out} \gg 1$.

This yields equivalent noise bandwidths

$$B_{Eq.,\,Hold} \cong \frac{G_m}{4C_L}, \quad (3.11)$$

and

$$B_{Eq.,\,Track} \cong \frac{G_m}{4(C_H + C_L)}. \quad (3.12)$$

Assuming that the OTA contains only CMOS transistors and thus exhibits thermal noise apart from $1/f$-noise, its equivalent input-referred voltage density for this type of noise is frequency independent and it is inversely proportional to its transconductance:

$$\frac{\overline{v_n^2}}{\Delta f} \sim \frac{1}{G_m}. \quad (3.13)$$

This yields a total wideband noise $\overline{v_n^2} \sim \frac{1}{4}\left(\frac{1}{C_H + C_L} + \frac{1}{C_L}\right)$. Hence the capacitors C_H and C_L play an eminent role in the noise performance of the THA.

Let us now have a look at the accuracy requirements. At the time t_{H1}, i.e. at the end of the hold period, we require an accuracy corresponding to n-bit, i.e. the compound error must be smaller than a half *LSB*. This consists of a deterministic error ε_{Det} and a random error due to noise. The relative error $\varepsilon_{Rel,\,Det}$ can be determined using the above equations, namely:

$$\varepsilon_{Rel,Det} = 1 - \left[\frac{G_m R_{Out}}{G_m R_{Out} + 1} + \frac{V_{OS}}{(G_m R_{Out} + 1) \cdot V_{In}(t_0)} \right] \cdot \left[1 - e^{-\frac{(G_m R_{Out} + 1)}{C_L R_{Out}} t_{H1}} \right], \tag{3.14}$$

for the hold phase, while the random error is due to the noise voltage $\overline{v_n^2}$ (see Eq. (3.13)). Hence we obtain for the hold phase:

$$\left| V_{Error,\ \max} \right| = \varepsilon_{Rel,Det,\ \max} \cdot \left| V_{In,\ \max} \right| + \sqrt{\overline{v_n^2}} \le \frac{\left| V_{In,\ \max} \right|}{2^{n+1}}, \tag{3.15}$$

where $V_{In}(t_0)$ in the above equation is replaced by $V_{In,\ \max}(t_0)$. While this equation shows that high transconductance and high output resistance R_{Out} of the OTA are of extreme importance, the role of C_H and C_L is ambivalent: for low deterministic error both capacitors should be small, but the noise requirements call for large capacitors. Obviously, the design of "good" THA requires a lot of optimization and some trade-offs are necessary.

Furthermore, any resistive loading at the OTA output would degrade the OTA performance and should be avoided.

Note that if we accept that the gain of the converter is not exactly unity, the factor $G_m R_{Out} / (G_m R_{Out} + 1)$ does not play a prominent role in our considerations. Nevertheless, if $G_m R_{Out} \gg 1$, we can write $\frac{G_m R_{Out} + 1}{C_L R_{Out}} \cong \frac{G_m}{C_L}$ which corresponds to $f_{3dB,\ Hold}$ and $\frac{G_m R_{Out} + 1}{(C_H + C_L) \cdot R_{Out}} \cong \frac{G_m}{C_H + C_L} = 2\pi\ f_{3dB,\ Track}$.

Then, we can write

$$\varepsilon'_{Rel,Det} = 1 - \left[1 - \frac{V_{OS}}{(G_m R_{Out} + 1) \cdot V_{In}(t_0)} \right] \cdot \left[1 - e^{-2\pi f_{3dB,\ Hold} \cdot t_{H1}} \right]. \tag{3.16}$$

Hence, for a given error $\varepsilon'_{Rel,Det}$ we can calculate the required OTA bandwidth as:

$$f_{3dB,\ Hold} \cong \frac{1}{2\pi \cdot t_{H1}} \cdot \ln \frac{1 - \xi}{\varepsilon'_{Rel,Det} - \xi}, \tag{3.17}$$

where

$$\xi = \frac{V_{OS}}{(G_m R_{Out} + 1) \cdot V_{In}(t_0)}. \tag{3.18}$$

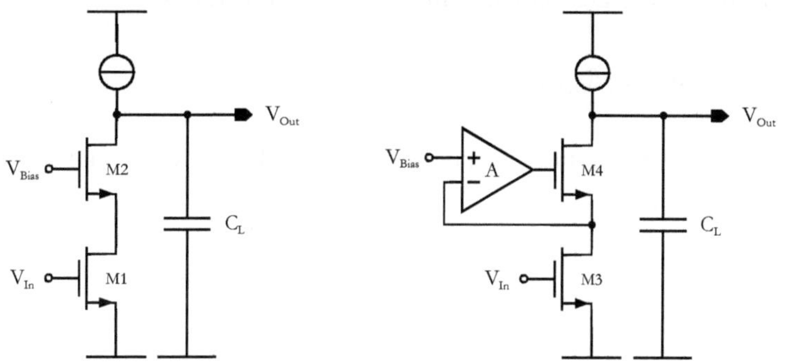

Fig. 3.7: Conventional cascode amplifier and active cascode amplifier

3.2.2 Folded-Cascode Operational Transconductance Amplifier

Due to the single amplification stage of the *folded-cascode operational transconductance amplifier*, the folded-cascode OTA offers the best trade-off between the speed and the accuracy for high-speed sampling [84]. Note that unity-gain bandwidth for OTA model shown above is $f_{0dB} = \frac{1}{2\pi} \cdot \frac{C_L R_{Out}}{G_m R_{Out} + 1}$ for $C_H = 0$.

Concerning to the THA performance, higher unity-gain frequency f_{0dB} and the use of a single gain stage with the dominant pole at the amplifier output lead to a faster settling process, whereas higher DC voltage gain (given by $G_m R_{Out}$ for the OTA) leads to more accurate settling. The folded-cascode OTA is basically a single gain stage amplifier. The higher DC voltage gain can be obtained by using *gain-boosting technique* (also known as *active cascode amplifier*) to reach the accuracy requirement [84-86].

A conventional *cascode amplifier* featuring a single-ended input is illustrated on the left side of Fig. 3.7. The output impedance R_{Out} of the conventional cascode amplifier can be described as [84]:

$$R_{Out} = R_{DS1} + R_{DS2} + g_{m2} \cdot R_{DS1} \cdot R_{DS2} \,, \tag{3.19}$$

where the resistances R_{DS1} and R_{DS2} represent the drain-source resistances of the transistors M1 and M2, respectively, and the transconductance g_{m2} represents the small-signal transconductance of the cascode transistor M2.

Furthermore, the *DC voltage gain* A_0 can be written as:

$$A_0 = g_{m,\,eff} \cdot R_{Out} \;, \tag{3.20}$$

whereas the *effective transconductance* $g_{m,\,eff}$ of the conventional cascode amplifier is equal to:

$$g_{m,\,eff} \equiv \frac{\Delta I}{\Delta V_{In}} = g_{m1} \cdot \frac{\left(g_{m2} R_{DS1} + \dfrac{R_{DS1}}{R_{DS2}} \right)}{\left(g_{m2} R_{DS1} + \dfrac{R_{DS1}}{R_{DS2}} + 1 \right)} \;. \tag{3.21}$$

Hence, the DC voltage gain A_0 can be expressed as:

$$A_0 = g_{m1} R_{DS1} \cdot \left(g_{m2} R_{DS2} + 1 \right) \;. \tag{3.22}$$

And the corresponding unity-gain frequency f_{0dB} is expressed to be:

$$f_{0dB} \cong \frac{1}{2\pi} \cdot \frac{C_L \cdot R_{Out}}{A_0 + 1} \;, \tag{3.23}$$

where the parameter C_L represents the *capacitive load* of the cascode amplifier in Fig. 3.7. Note that this neglects the second (i.e. nondominant) pole at the source of the cascode device M2. Increasing the output resistance R_{Out} of the conventional cascode amplifier is the best way to enhance the DC voltage gain A_0 without reducing the unity-gain frequency f_{0dB} (then $f_{0dB} \cong \frac{1}{2\pi} \cdot \frac{C_L}{G_m}$).

A modified cascode amplifier, suggested by [84-86], is illustrated on the right side of Fig. 3.7. Such modified cascode amplifier is known in the literature as the *active cascode amplifier* or simply as the *gain-boosting stage*.

The output resistance R_{Out} of the active cascode amplifier is substantially increased and can be written as [84]:

$$R_{Out} = R_{DS1} + R_{DS2} + g_{m2} \cdot R_{DS1} \cdot R_{DS2} \cdot (A+1) \;, \tag{3.24}$$

where A is the gain of the amplifier inserted between the gate and the source of the transistor M4 in Fig. 3.7.

The effective transconductance $g_{m,\,eff}$ of the active cascode amplifier changes very little and can be described as:

$$g_{m,\,eff} \equiv \frac{\Delta I}{\Delta V_{In}} = g_{m1} \cdot \frac{\left(g_{m2} R_{DS1} \cdot (A+1) + \frac{R_{DS1}}{R_{DS2}} \right)}{\left(g_{m2} R_{DS1} \cdot (A+1) + \frac{R_{DS1}}{R_{DS2}} + 1 \right)} . \tag{3.25}$$

But the total DC voltage gain A_0 of the active cascode amplifier is improved significantly due to high R_{Out} and can be described as:

$$A_0 = g_{m1} R_{DS1} \cdot \left(g_{m2} R_{DS2} \cdot (A+1) + 1 \right) . \tag{3.26}$$

Fig. 3.8 depicts a folded-cascode operational transconductance amplifier designed with the gain-boosting technique (transistors M5 ~ M10) to increase the total DC voltage gain A_0 of the operational transconductance amplifier. The input differential pair, M1 and M2, is composed of *MOS transistors of depletion type* to enhance the common-mode input range $\left| V_{In,\,CM} \right|$ of the amplifier.

At the beginning of analysis, considering at first the situation without the gain-boosting interstage, M5 ~ M10, and the *Miller's capacitor* C_M , the small-signal voltage gain of this amplifier at low frequencies can be described as [86]:

$$A_0 = G_m \cdot R_{Out} , \tag{3.27}$$

where the parameter G_m is the transconductance of the amplifier and R_{Out} is its output resistance. The transconductance G_m is equal to the transconductance of the input differential pair M1 and M2, i.e. g_{m1} and g_{m2} . Besides, the output resistance R_{Out} can be written as the parallel connection of the output resistance $R_{Out,\,M14}$ and $R_{Out,\,M16}$:

$$R_{Out} = R_{Out,\,M14} \,||\, R_{Out,\,M16} , \tag{3.28}$$

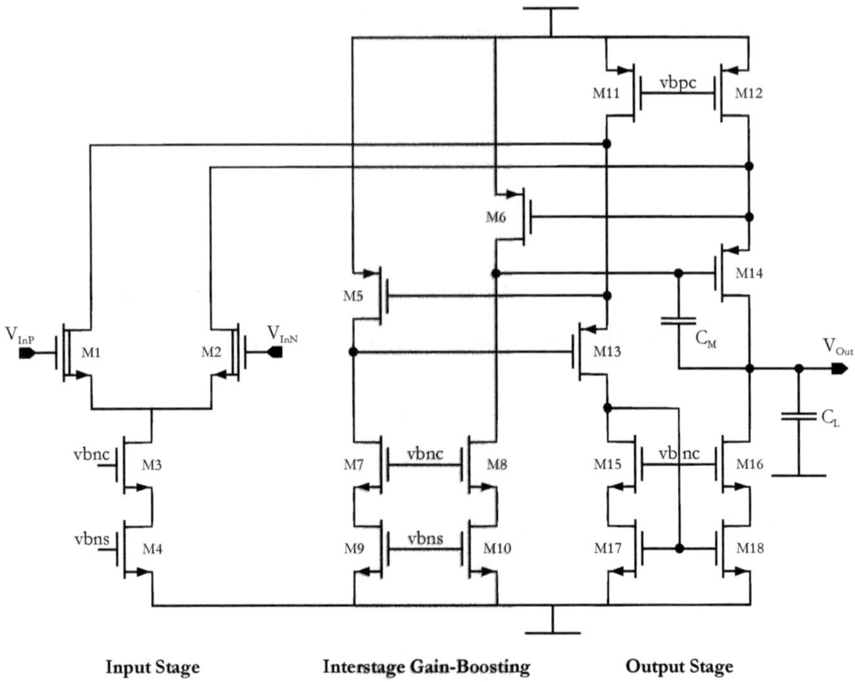

Fig. 3.8: A folded-cascode OTA with interstage gain-boosting

where the output resistance $R_{Out,M14}$ and $R_{Out,M16}$, which are observed at the drain terminals of the transistors M14 and M16 can be expressed as [86]:

$$\begin{aligned} R_{Out,M14} &= \left(R_{DS2} \parallel R_{DS12}\right) + R_{DS14}\left[1 + g_{m14} \cdot \left(R_{DS2} \parallel R_{DS12}\right)\right] \\ &\cong g_{m14} \cdot R_{DS14} \cdot \left(R_{DS2} \parallel R_{DS12}\right), \end{aligned} \tag{3.29}$$

and

$$\begin{aligned} R_{Out,M16} &= R_{DS18} + R_{DS16}\left(1 + g_{m16} \cdot R_{DS18}\right) \\ &\cong g_{m16} \cdot R_{DS16} \cdot R_{DS18}, \end{aligned} \tag{3.30}$$

respectively. Note that R_{DSi} is the drain-source resistance of the transistor Mi.

Let us now consider the effect of the gain-boosting interstage on the total DC voltage gain A_0. Instead of an amplifier composed of a differential pair as illustrated on the right side of Fig. 3.7, *common-source amplifiers* containing M5 or M6 are used as the amplifiers in the gain-boosting stage. As an example, the voltage gain A of the common-source amplifier M5 can be summarized as [87, 88]:

$$A = g_{m5} \cdot \left[R_{DS5} \parallel \left(R_{DS7} + R_{DS9} \right) \right] . \tag{3.31}$$

While the output resistance $R_{Out,M16}$ does not change, but the output resistance $R_{Out,M14}$ is increased correspondingly and can be expressed as:

$$\begin{aligned} R_{Out,M14} &= (R_{DS2} \parallel R_{DS12}) + R_{DS14} \left[1 + g_{m14} \cdot (A+1) \cdot (R_{DS2} \parallel R_{DS12}) \right] \\ &\cong g_{m14} \cdot R_{DS14} \cdot (A+1) \cdot (R_{DS2} \parallel R_{DS12}) . \end{aligned} \tag{3.32}$$

But, while the output resistance $R_{Out,M14}$ and $R_{Out,M16}$ are connected in parallel, the interstage gain-boosting has to be also applied to the output resistance $R_{Out,M16}$ to increase the total output resistance R_{Out}.

The Miller's capacitor C_M in Fig. 3.8 improves the *phase margin* Φ_M and, hence, contributes to the frequency compensation.

3.3 Subsystem: Fine Converter

3.3.1 Introduction and Definitions

We now turn out attention to the fine converter (see Fig. 2.19). The aim of the fine converter inside the folding and interpolating A/D converter is to process the input signal, either a sampled or an unsampled signal, and to resolve it into the lower significant bits of the word length of the whole analog-to-digital converter. The output signals of the fine converter are then synchronized with output signals of the coarse converter and combined to yield the output of the complete converter. Using such arrangement can reduce the number of needed comparators immensely, when compared with the flash A/D converter.

The fine converter is composed of a folding stage, an interpolation stage, a comparator stage, and an error correction stage. The arrangement can be seen in the block diagram illustrated in Fig. 3.9. The details of each stage will be treated in the following sections.

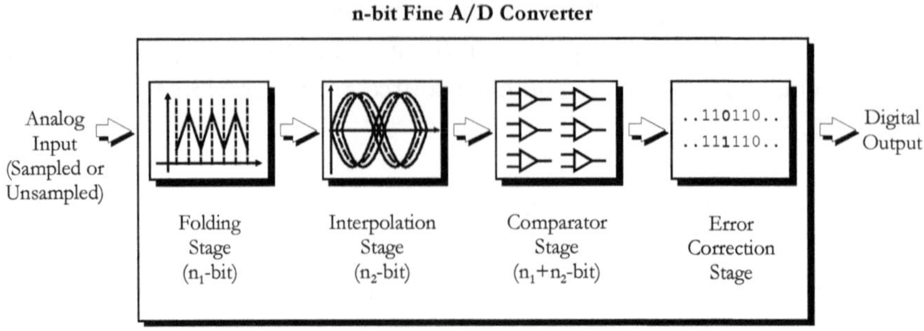

Fig. 3.9: Block diagram of the fine converter

For an n-bit fine converter, this resolution n can be split into two parts, namely n_1 and n_2. The first number n_1 indicates the contribution of resolution from the folding stage and the second number n_2 indicates the contribution due to interpolation. The mathematical relationships among these resolutions are summarized as follows:

$$n = n_1 + n_2 \,, \tag{3.33}$$

where

$$n_1,\ n_2 \in \mathbb{N}, \quad 1 \le n_1 \le n \quad \text{and} \quad 0 \le n_2 \le n-1 \,. \tag{3.34}$$

The extreme situations, $n_1 = n$ and $n_2 = 0$, meaning that the interpolation stage vanishes and the resolution of the fine converter is determined only by the folding stage.

For the convenience of discussion, the terminology used for the folding and interpolating A/D converter with m-bit coarse converter and n-bit fine converter is summarized below. The definitions and interpretations will be treated in upcoming sections.

Terminology of resolution:

System resolution:

$$A_{Res} = m + n = m + (n_1 + n_2) \,. \tag{3.35}$$

Folding factor (*coarse resolution*):

$$F_F = 2^m \,. \tag{3.36}$$

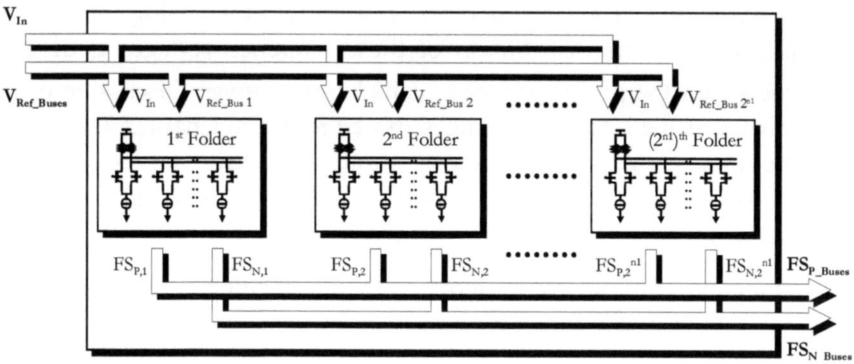

Fig. 3.10: Structure of the folding stage

Number of folders (also known as *number of folding blocks,* abbreviated as FB or *folding resolution*):

$$N_{FB} = \frac{N_{FS}}{F_{ITPL}} = \frac{2^{n_1+n_2}}{2^{n_2}} = 2^{n_1} \ . \tag{3.37}$$

Interpolating factor (*interpolating resolution*):

$$F_{ITPL} = 2^{n_2} \ . \tag{3.38}$$

Number of folding signals (*fine resolution*):

$$N_{FS} = 2^{n} = 2^{n_1+n_2} \ . \tag{3.39}$$

Number of totally representable states:

$$2^{(m+n)} = F_F \cdot N_{FS} = F_F \cdot (N_{FB} \cdot F_{ITPL}) = 2^m \cdot 2^{n_1} \cdot 2^{n_2} \ . \tag{3.40}$$

3.3.2 Topology and Functions of the Folding Stage

The folding stage is primarily composed of several *folders* (also named *folding blocks,* FBs), which are connected in parallel. An example of a folding stage is illustrated in Fig. 3.10. The analog input signal bus V_{In} is fed to all folders in parallel. It can be either a sampled or an unsampled signal, as discussed above. The other inputs are the bus signals V_{Ref_Buses} that correspond to fixed predefined reference voltages, which are distributed to the folders in a particular arrangement in order to fold the analog input signal V_{In} equidistantly. The distribution of these reference voltages is of great consequence for the positioning of so called *zero-crossings*. Discussion of this arrangement of the reference voltages can be found in Chapter 3.3.3. Apart from the folders, components such as amplifiers to drive the next stage or resistors for *averaging technique* can also be contained in this stage. The effect of *averaging resistors* is treated in the next section.

The function of a folder is to separate the input signal range into several intervals. The output signal, either in form of voltage signal or in form of current signal, of the folder is called a *folding signal* (FS). The folding signals are differential signals; therefore, they are generally marked with a sign to indicate the pole direction. The number of folders N_{FB} represents the resolution which the folder stage can contribute, for example n_1 as shown in Fig. 3.10. Consequently, the index N_{FB} can be usually described in terms of power of 2.

3.3.3 Topology and Functions of the Folding Block

Fig. 3.11 shows the structure of the jth folder in the folding stage. The folder itself contains a certain number of *folding amplifiers* (FA), output signals of which are cross-connected in order to generate folded signals. Because the output signals of folding amplifiers are cross-connected, the loads of all folding amplifiers can be combined together as in Fig. 3.11, if that is necessary.

The number of folding amplifiers is mainly determined by the folding factor F_F, which is an even number and equal to 2^m (see Eq. (3.36)). Because the operating points of the transistors at both ends of the input signal range are near to the edge of the *triode region*, two extra folding amplifiers, the reference voltages of which are outside both ends of the input signal range, are used to compensate signal distortion at both ends. Since their reference voltages are outside the input signal range, for the convenience of discussion, these two folding amplifiers will be neglected in the following text. Furthermore, a DC offset is added to the folding signal FS_N to generate an offset level shifting to ensure the total number of folding amplifiers to be an odd number. Hence, only 2^m folding amplifiers and this DC offset are considered in Fig. 3.11.

As discussed above, the folder has two buses as input signals: V_{In} and V_{Ref_Buses}. Both inputs are connected in parallel to the folding amplifiers. The folding amplifier is usually realized as a differential amplifier. The ref-

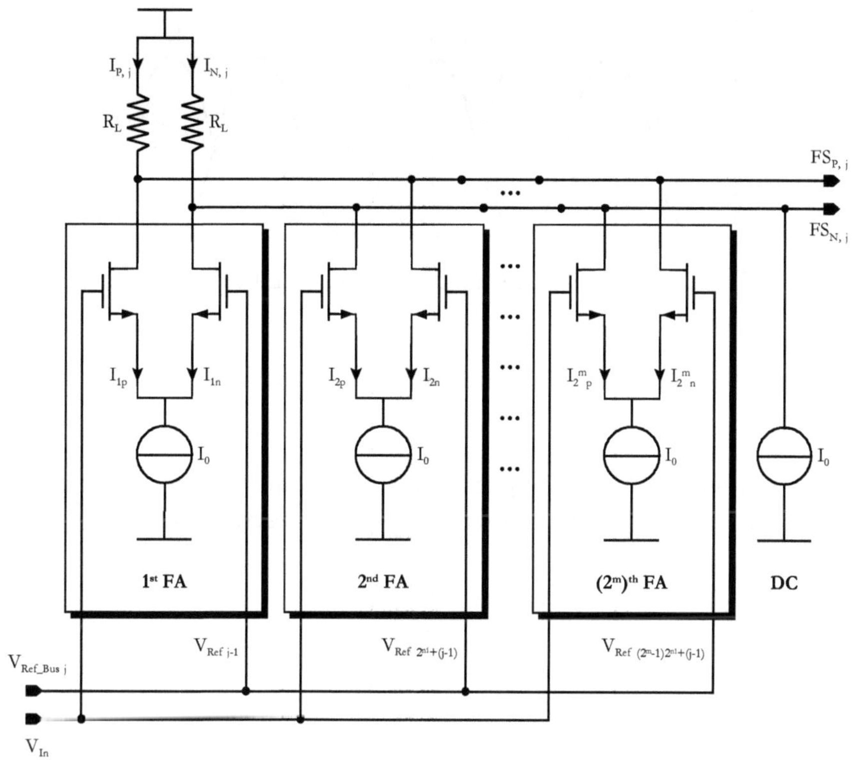

Fig. 3.11: Structure of the jth folder in the folding stage

erence voltage of the kth folding amplifier of the jth folder in the folding stage can be described as $V_{Ref\ (k-1)\cdot 2^{m1}+(j-1)}$. Thus, when the input signal V_{In} is equal to this reference voltage, the currents I_{kp} and I_{kn} of the kth folding amplifier are equal to the half of I_0,

$$I_{kp} = I_{kn} = \frac{I_0}{2} . \tag{3.41}$$

Since the remaining number of folding amplifiers is now an even number and the outputs of all folding amplifiers are cross-connected with every even amplifier with reversed output polarity, the differential voltage at the folder output exhibits zero-crossings at the reference voltages. But only if the polarity of every other folding amplifier differential output has been connected to the common output in reverse manner because then all output currents can-

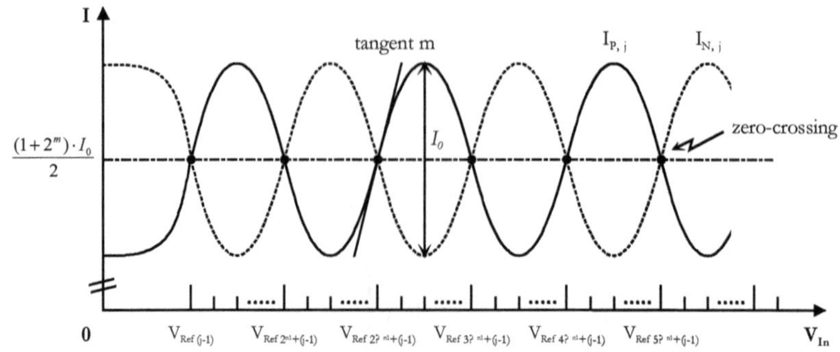

Fig. 3.12: Positioning of folding signals and zero-crossings

cel except those due to the amplifier operating near the zero-crossing. The currents $I_{P,j}$ and $I_{N,j}$ at the zero-crossing can be described as:

$$I_{P,j} = I_{N,j} = \frac{I_0}{2} + \frac{2^m \cdot I_0}{2} = \frac{(1+2^m)\cdot I_0}{2} . \qquad (3.42)$$

General form of the load currents $I_{P,j}$ and $I_{N,j}$ can be written as:

$$I_{P,j} = I_{1p} + I_{2n} + I_{3p} + I_{4n} + \dots + I_{(2^m-1)p} + I_{2^m n} , \qquad (3.43)$$

and

$$I_{N,j} = I_{1n} + I_{2p} + I_{3n} + I_{4p} + \dots + I_{(2^m-1)n} + I_{2^m p} + I_0 . \qquad (3.44)$$

An example of the differential folding signal and its corresponding zero-crossings as a function of the input voltage V_{In} is illustrated in Fig. 3.12. If the tangent m corresponding to transconductance of the folding amplifier is well defined, then the signal level of folding signals can reach its local maxima and minima as indicated in the figure. Supposing that the folding signals in Fig. 3.12 are output signals of the jth folder in the folding stage, the first zero-crossing is located at $V_{Ref\ (j-1)}$ and the second one is positioned at $V_{Ref\ 1 \cdot 2^{n1}+(j-1)}$, as depicted in Fig. 3.13. The following zero-crossings are positioned equidistantly.

Considering again the plot shown in Fig. 3.12, it can be observed that the "input-signal-based" characteristic is very similar to a "time-based" periodical signal. Such kind of signal is also named as *Gilbert's sine generator* [47, 70].

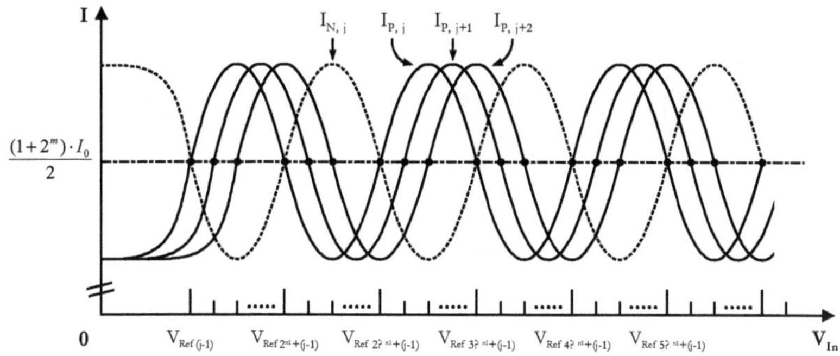

Fig. 3.13: Distribution of shifted folding signals

Since the loads of the folding amplifier are simple resistors (Fig. 3.11), the current signals $I_{P,j}$ and $I_{N,j}$ in Fig. 3.12 cause that the voltage signals $FS_{P,j}$ and $FS_{N,j}$ at the output (see Fig. 3.11) exhibit the same waveform. Providing that all reference voltages used to create the folding signals in Fig. 3.12 are shifted to the next reference voltage on the right hand side, the positioning of zero-crossings is also shifted to the right hand side correspondingly. Strictly speaking, this new folding signal pair FS_{j+1} is generated from the $(j+1)^{th}$ folder in Fig. 3.10.

As an example, the respective folding signals can be found in Fig. 3.13. The newly generated current signals, $I_{P,j+1}$ and $I_{P,j+2}$, seem to be phase-shifted signals with respect to the first folding signal $I_{P,j}$. But, actually the x-axis is a sweep of the input signal, not of the time. In this figure, it can be also seen that the current folding signal $I_{N,j}$ has the same position as $I_{P,j+4}$. This characteristic is an issue of major importance in the signal interpolation, which will be treated in the later section.

The folding factor F_F is a factor, which can be chosen by the designer. Changing the folding factor F_F has an impact both on both, the coarse and fine converters. As a direct definition, the folding factor F_F is the total number of representable states of the coarse converter. To put it differently, it describes how many times the input signal has been folded to cover the input range for the fine converter, hence the name "folding factor". The definition of the folding factor F_F can be found in Eq. (3.6) and equal to 2^m.

The input reference voltages available at corresponding buses of the folding stage mentioned in the beginning in this section (Fig. 3.10) can now be recapitulated and listed as below.

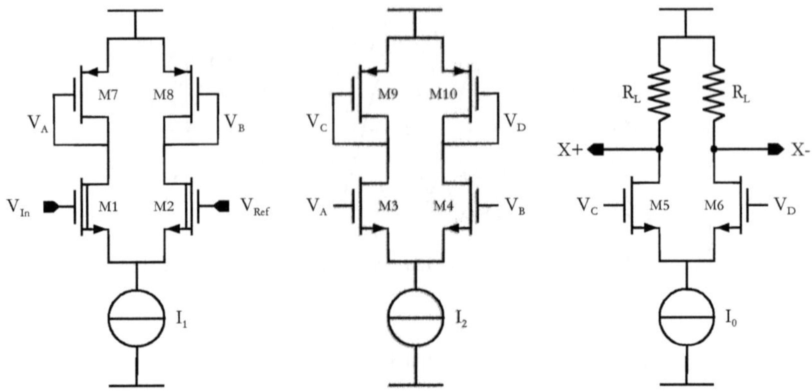

Fig. 3.14: Cascaded amplifiers as folding amplifier

For the 1st folder, the reference voltages are:

$$V_{Ref\_Bus1} = \{V_{Ref\ 0},\ V_{Ref\ (1\cdot 2^{n1})},\ V_{Ref\ (2\cdot 2^{n1})},\ \cdots,\ V_{Ref\ [(2^m-1)\cdot 2^{n1}]}\}\ . \tag{3.45}$$

For the 2nd folder, the reference voltages can be expressed as follows:

$$V_{Ref\_Bus2} = \{V_{Ref\ 1},\ V_{Ref\ (1\cdot 2^{n1}+1)},\ V_{Ref\ (2\cdot 2^{n1}+1)},\ \cdots,\ V_{Ref\ [(2^m-1)\cdot 2^{n1}+1]}\}\ . \tag{3.46}$$

And, for the last folder, namely the (2^{n_1}) th folder, its corresponding reference voltages are:

$$\begin{aligned} V_{Ref\_Bus2^{n1}} &= \left\{ \begin{matrix} V_{Ref\ (2^{n1}-1)}, V_{Ref\ [1\cdot 2^{n1}+(2^{n1}-1)]}, \\ V_{Ref\ [2\cdot 2^{n1}+(2^{n1}-1)]}, \ldots, V_{Ref\ [(2^m-1)\cdot 2^{n1}+(2^{n1}-1)]} \end{matrix} \right\} \\ &= \left\{ \begin{matrix} V_{Ref\ (2^{n1}-1)}, V_{Ref\ [1\cdot 2^{n1}+(2^{n1}-1)]}, \\ V_{Ref\ [2\cdot 2^{n1}+(2^{n1}-1)]}, \ldots, V_{Ref\ (2^{m+n1}-1)} \end{matrix} \right\}. \end{aligned} \tag{3.47}$$

Hence, the total number of reference voltages to be generated for the folding and interpolating A/D converter is 2^{m+n_1}, which is 2^{n_2}-times smaller than the flash A/D converter with the same resolution.

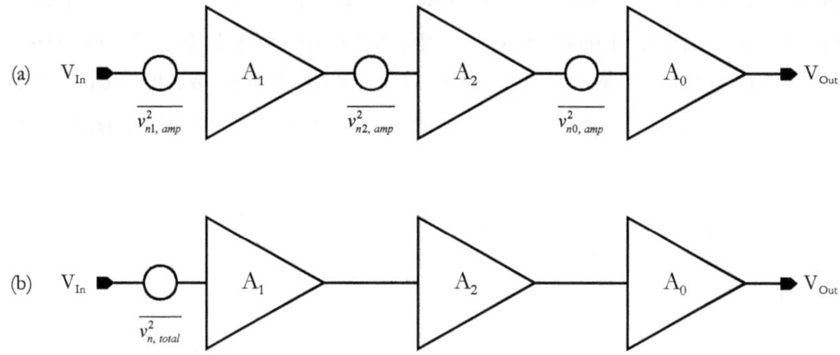

Fig. 3.15: Noise of cascaded amplifiers

3.3.4 Topology and Circuit Design of the Folding Amplifier

The *folding amplifier* is the basic component of the folder. Actually, the folding amplifier can be realized using a simple differential amplifier. Only the outputs of these folding amplifiers must be cross-connected, i.e. the polarity of every other folding amplifier differential output has to be connected in reverse manner, as shown in Fig. 3.11.

The small signal analysis for a single differential amplifier with active load is treated in Appendix A. According to the analysis in the appendix, the small signal differential-mode voltage gain of a folding amplifier implemented employing a simple differential amplifier, as depicted in Fig. 3.11, can be described as below:

$$A_{DM} \cong -\frac{g_m}{g_L}, \tag{3.48}$$

for the amplifier with active PMOS load and

$$A \cong -g_m \cdot R_L, \tag{3.49}$$

for amplifiers with resistive load R_L.

The transconductance g_m (and also the output conductance g_{DS} and thus the conductance g_L) depend on channel length of the MOS transistor employed. However, the *channel length* L is normally constrained by the fabrication process and it cannot be shorter than the limit for a given technology to obtain a higher g_m. Second, the load is shared with other folding amplifiers to set the output signal range of the folding signal (Fig. 3.11). Thus, the best way to increase the voltage gain A_{DM} is to choose a MOS device at the input that features large channel width W to obtain a high transconductance.

Another approach to implement the folding amplifiers (FA) featuring high voltage gain is to use *cascaded amplifiers* instead of a single amplifier as indicated in Fig. 3.14. The amplification factor for differential signals of each stage, A_1, A_2, and A_0, can be described as follows:

$$A_1 \cong -\frac{g_{m1}}{g_{m7}}, \tag{3.50}$$

$$A_2 \cong -\frac{g_{m3}}{g_{m9}}, \tag{3.51}$$

and

$$A_0 \cong -g_{m5} \cdot R_L. \tag{3.52}$$

The total voltage gain A can be deduced from multiplication of all three stages. This becomes:

$$A = A_1 \cdot A_2 \cdot A_0, \tag{3.53}$$

$$\Rightarrow A \cong -\frac{g_{m1} \cdot g_{m3} \cdot g_{m5} \cdot R_L}{g_{m7} \cdot g_{m9}}. \tag{3.54}$$

In order to obtain high total voltage gain A, there are now two possibilities. One is to increase the gate width of input transistors, M1~M6. The other one is to decrease the gate width of the active loads, namely M7~M10.

The disadvantage of such topology shown in Fig. 3.14 is that it increases the noise by the cascaded amplification (see Fig. 3.15). Supposing that the *input-referred equivalent noise* of the amplifiers, i.e. A_1, A_2, and A_0, can be described as $\overline{v_{n1,\,amp}^2}$, $\overline{v_{n2,\,amp}^2}$, and $\overline{v_{n0,\,amp}^2}$, respectively, the total input-referred equivalent noise of the cascaded amplifier chain $\overline{v_{n,\,total}^2}$ can be written as:

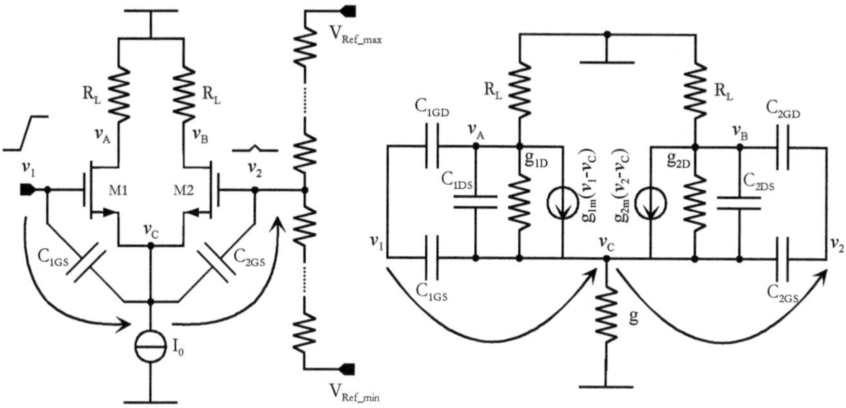

Fig. 3.16: Small signal analysis for input feedthrough

$$\overline{v_{n,\,total}^2} = \overline{v_{n1,\,amp}^2} + \frac{\overline{v_{n2,\,amp}^2}}{A_1^2} + \frac{\overline{v_{n0,\,amp}^2}}{A_1^2 \cdot A_2^2} \,. \tag{3.55}$$

Hence, the noise of the 1st amplifier is dominating the performance of the cascaded amplification.

The input-referred equivalent noise of the amplifiers A_1, A_2, and A_0 can be expressed as follows:

$$\overline{v_{n1,\,amp}^2} = 2\left[\overline{v_{n1}^2} + \overline{v_{n7}^2} \cdot \left(\frac{g_{m7}}{g_{m1}}\right)^2\right] \quad \text{for } g_{m1} = g_{m2} \text{ and } g_{m7} = g_{m8}, \tag{3.56}$$

$$\overline{v_{n2,\,amp}^2} = 2\left[\overline{v_{n3}^2} + \overline{v_{n9}^2} \cdot \left(\frac{g_{m9}}{g_{m3}}\right)^2\right] \quad \text{for } g_{m3} = g_{m4} \text{ and } g_{m9} = g_{m10}, \tag{3.57}$$

$$\overline{v_{n0,\,amp}^2} = 2\left[\overline{v_{n5}^2} + \frac{\overline{v_{n,\,R_L}^2}}{g_{m5}^2 \cdot R_L^2}\right] \quad \text{for } g_{m5} = g_{m6}\,. \tag{3.58}$$

For the input-referred equivalent noise of the transistor M_i it is valid

$$\frac{\overline{v_{ni}^2}}{\Delta f} = \frac{k'}{C_{OX} \cdot W_i \cdot L_i \cdot f} + 4kT \cdot \left(\frac{2}{3g_{mi}} \right), \tag{3.59}$$

while the thermal noise of the resistor R_L is

$$\frac{\overline{v_{n,\,R_L}^2}}{\Delta f} = 4kT \cdot R_L \ . \tag{3.60}$$

Consequently, for $g_{m1}/g_{m7} \gg 1$ we obtain $\overline{v_{n,\,total}^2} \cong 2 \cdot \overline{v_{n1}^2}$.

For an $(m+n)$-bit folding and interpolating A/D converter, the folding stage in the n-bit fine converter still has to reach $(m+n)$-bit accuracy (see Fig. 2.19). Thus, the total input-referred equivalent noise $\overline{v_{n,\,total}^2}$ has to be limited to:

$$\overline{v_{n,\,total}^2} \le \frac{LSB}{2} = \frac{|V_{In\_\max}|}{2^{m+n+1}} \ . \tag{3.61}$$

Applying the averaging technique on the output signals can reduce this noise. For example, the wires A, B, C and D can be connected to the same wires in another folder, which reference voltages are adjacent to the averaged folder. Details about the averaging techniques are treated in the later section in this chapter.

In order to obtain maximum voltage gain from the differential pairs, the transistors must operate in saturation region. This leads to restriction on the output voltage swing. For example, the output signal range of one amplifier should fit inside the input signal range of the next amplifier.

In addition, the first input transistor pair, M1 and M2, contains *depletion MOS transistors* to ensure large input voltage range of the folder. The restriction on the input signal range can be found by investigating the behavior of the differential amplifier, as discussed in the following.

Besides the common performance parameters, such as *differential-mode voltage gain* A_{DM}, *common-mode voltage gain* A_{CM}, and the *common-mode rejection ratio* (CMRR), *noise, bandwidth* etc., there are other parameters of high importance. In our application we have found that the *input feedthrough* at the input transistor pair is also important. In Fig. 3.16, a basic MOS differential amplifier with a load resistor R_L is depicted, in association with its small-signal equivalent circuit diagram. Assuming that there is a rapid volt-

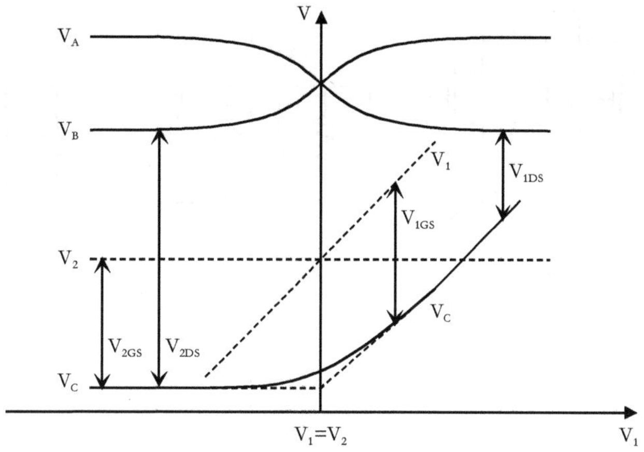

Fig. 3.17: DC transfer characteristics of the differential amplifier

age transient v_1 at one of its inputs and the other input is connected to constant reference voltage v_2 generated by a resistor string serving as a voltage divider and driven by a reference voltage source, this rapid signal change of v_1 side causes an interference due to capacitive feedthrough via C_{1GS} and C_{2GS}. The effect of this coupling depends on characteristics of the voltage sources connected to the amplifier inputs. If the reference voltage v_2 is not buffered and merely generated by the resistance string as shown in Fig. 3.16, the effect of the input feedthrough will cause disturbances of the reference voltage.

Considering the DC transfer characteristic of the differential amplifier, the voltages V_A, V_B, and V_C are plotted in Fig. 3.17 for V_2 staying constant and V_1 being swept across the input voltage range [89].

The signals V_A and V_B represent the output signals of the differential amplifier. These signals are equal when the input signal V_1 is equal to the reference signal V_2. What is important is the behavior of the voltage at the common mode node V_C. If the input signal V_1 is much lower than the reference signal V_2, the drain current of M1 is zero and the device M2 conducts all current of the current source I_0 (Fig. 3.16). Since the voltage V_2 and the current I_0 are constant, the common source voltage V_C is kept at constant potential due to a constant voltage V_{2GS}, as shown on the left-hand-side of Fig. 3.17. In this case, the voltage V_{2DS} is high enough to keep the transistor M2 in saturation region.

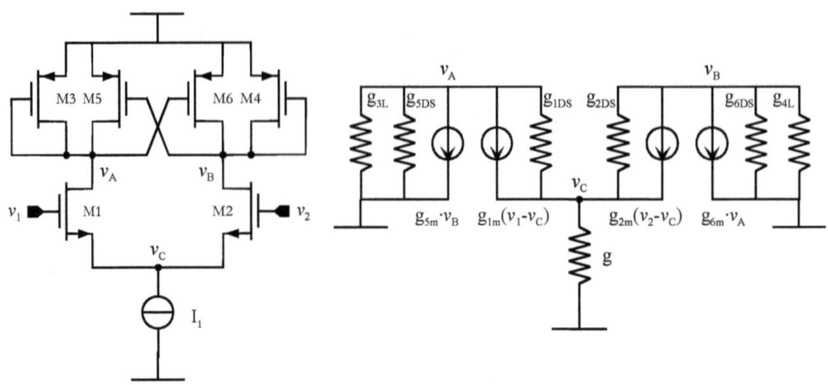

Fig. 3.18: Differential amplifier with "negative" conductance M5 and M6

Supposing that the input signal V_1 is now higher than the reference voltage V_2, then the voltage V_{1GS} increases and the voltage V_{2GS} decreases, as shown at the right-hand-side of Fig. 3.17. The drain current of M1 increases. Since the input signal V_1 increases along with the x-axis, the common source voltage V_C also increases.

If V_1 is sufficiently higher than V_2, the device M1 conducts all current I_0 and the device current of M2 is zero. Then V_{1GS} is constant and V_C rises at the same rate as V_1. A further rise of the common source voltage V_C leads to a decrease of voltage V_{1DS}, which can lead the device M1 into the triode region.

Considering fast transients of the input signal V_1, there are large changes in both voltages V_{1GS} and V_{2GS} and the capacitors C_{1GS} and C_{2GS} conduct displacement currents. Therefore, the voltage V_2 can be affected by capacitive feedthrough since the voltage V_2 is generated by a nonideal voltage source. How this feedthrough affects V_2, thus depends on the resistance of the resistor string and values of C_{1GS} and C_{2GS}.

An idea to improve the differential amplifier is to replace the load with two pairs of transistors M3 and M4, M5 and M6, as depicted in Fig. 3.18. In Fig. 3.18, the transistor pair M3 and M4 works as active load and the transistor pair M5 and M6 is a cross-coupled current mirror.

The equivalent circuit diagram for the small-signal behavior analysis is displayed on the left side of Fig. 3.18. According to the Kirchhoff's law and the principle of superposition, the branch point voltages v_A, v_B, and v_C can be given as follows:

$$(g_{3L}+g_{5DS})\cdot v_A+g_{5m}\cdot v_B+ \\ g_{1m}\cdot(v_1-v_C)+g_{1DS}\cdot(v_A-v_C)=0\ , \tag{3.62}$$

$$(g_{4L}+g_{6DS})\cdot v_B+g_{6m}\cdot v_A+ \\ g_{2m}\cdot(v_2-v_C)+g_{2DS}\cdot(v_B-v_C)=0\ , \tag{3.63}$$

and

$$-g_{1m}\cdot(v_1-v_C)-g_{2m}\cdot(v_2-v_C)+ \\ g_{1DS}\cdot(v_C-v_A)+g_{2DS}\cdot(v_C-v_B)+g\cdot v_C=0\ . \tag{3.64}$$

Supposing that the transistors M1 and M2, M3 and M4, and M5 and M6 match ideally, i.e. they meet the following conditions,

$$g_{1m}=g_{2m}\ , \tag{3.65}$$

$$g_{1DS}=g_{2DS}\ , \tag{3.66}$$

$$g_{3L}\equiv g_{3m}+g_{3DS}=g_{4m}+g_{4DS}\equiv g_{4L}\ , \tag{3.67}$$

$$g_{5m}=g_{6m}\ , \tag{3.68}$$

$$g_{5DS}=g_{6DS}\ . \tag{3.69}$$

Eqs. (3.62) ~ (3.64) can be simplified as:

$$(g_{3L}+g_{5DS})\cdot v_A+g_{5m}\cdot v_B+ \\ g_{1m}\cdot(v_1-v_C)+g_{1DS}\cdot(v_A-v_C)=0\ , \tag{3.70}$$

$$(g_{3L}+g_{5DS})\cdot v_B+g_{5m}\cdot v_A+ \\ g_{1m}\cdot(v_2-v_C)+g_{1DS}\cdot(v_B-v_C)=0\ , \tag{3.71}$$

and

$$-g_{1m}\cdot(v_1-v_C)-g_{1m}\cdot(v_2-v_C)+$$
$$g_{1DS}\cdot(v_C-v_A)+g_{1DS}\cdot(v_C-v_B)+g\cdot v_C=0\,. \tag{3.72}$$

From Eq. (3.72), the voltage v_C can be expressed as:

$$v_C=\frac{g_{1m}\cdot(v_1+v_2)+g_{1DS}\cdot(v_A+v_B)}{(g+2g_{1m}+2g_{1DS})}\,. \tag{3.73}$$

The term (v_A+v_B) in Eq. (3.73) can be deduced from summing Eq. (3.70) and Eq. (3.71).

$$(v_A+v_B)=\frac{2(g_{1m}+g_{1DS})\cdot v_C-g_{1m}\cdot(v_1+v_2)}{(g_{1DS}+g_{3L}+g_{5DS}+g_{5m})}\,, \tag{3.74}$$

so that

$$v_C=\frac{(g_{1DS}+g_{3L}+g_{5DS}+g_{5m})\cdot(v_A+v_B)+g_{1m}\cdot(v_1+v_2)}{2(g_{1m}+g_{1DS})}\,. \tag{3.75}$$

Inserting Eq. (3.74) into Eq. (3.73) and solving for the voltage v_C leads to:

$$v_C=\frac{g_{1m}\cdot(g_{3L}+g_{5DS}+g_{5m})\cdot(v_1+v_2)}{g\cdot g_{1DS}+(g+2g_{1m}+2g_{1DS})\cdot(g_{3L}+g_{5DS}+g_{5m})}$$
$$=f(v_1,v_2)\,, \tag{3.76}$$

which shows that the voltage v_C is function of the input signals v_1 and v_2.

The common-mode voltage gain A_{CM} can be calculated by applying Eq. (3.73) and Eq. (3.75) as follows:

$$\frac{g_{1m}\cdot(v_1+v_2)+g_{1DS}\cdot(v_A+v_B)}{(g+2g_{1m}+2g_{1DS})}$$
$$=\frac{(g_{1DS}+g_{3L}+g_{5DS}+g_{5m})\cdot(v_A+v_B)+g_{1m}\cdot(v_1+v_2)}{2(g_{1m}+g_{1DS})}\,. \tag{3.77}$$

This yields:

$$A_{CM} \equiv \frac{(v_A + v_B)}{2} \cdot \frac{2}{(v_1 + v_2)} = \frac{(v_A + v_B)}{(v_1 + v_2)}$$
$$= \frac{-g \cdot g_{1m}}{g \cdot g_{1DS} + (g + 2g_{1m} + 2g_{1DS}) \cdot [g_{3L} + (g_{5m} + g_{5DS})]} \, . \tag{3.78}$$

The term $(g_{5m} + g_{5DS})$ in the denominator of A_{CM} reduces the common-mode voltage gain A_{CM}. Considering the transistors M3, M4, M5 and M6 are designed to have the same transistor geometry (i.e. W and L) and the transconductance g_m is much more larger than the conductances g_{DS} and g, the common-mode voltage gain A_{CM} can be written as:

$$A_{CM} \cong \frac{-g \cdot g_{1m}}{g \cdot g_{1DS} + (g + 2g_{1m}) \cdot [g_{3m} + g_{5m}]}$$
$$\cong \frac{-g \cdot g_{1m}}{g \cdot g_{1DS} + 4g_{1m} \cdot g_{3m}} \cong \frac{-g}{4g_{3m}} \, . \tag{3.79}$$

The differential-mode voltage gain A_{DM} can be derived using a similar treatment, namely subtracting the two Eqs. (3.70) and (3.71). This leads to:

$$(v_A - v_B) = \frac{-g_{1m} \cdot (v_1 - v_2)}{(g_{1DS} + g_{3L} + g_{5DS} - g_{5m})} \, , \tag{3.80}$$

and

$$A_{DM} \equiv \frac{(v_A - v_B)}{(v_1 - v_2)} = \frac{-g_{1m}}{[g_{1DS} + g_{3L} - (g_{5m} - g_{5DS})]} \, . \tag{3.81}$$

The term $-(g_{5m} - g_{5DS})$ affects A_{DM}: since g_{5m} is normally larger than g_{5DS}, this term increases the differential-mode voltage gain A_{DM} of the amplifier circuit due to its negative sign. Because the sign of this term is negative, it acts like a "negative" resistance [71], [72], which is due to positive feedback of the devices M5 and M6 [90]. As discussed above, if the transistors M3, M4, M5, and M6 are designed to have the same geometry, the differential-mode voltage gain A_{DM} can be written as follows:

$$A_{DM} = \frac{-g_{1m}}{[g_{1DS} + g_{3L} - (g_{5m} - g_{5DS})]}$$
$$\cong \frac{-g_{1m}}{[g_{1DS} + (g_{3m} + g_{3DS}) - (g_{5m} - g_{5DS})]} \tag{3.82}$$
$$\cong \frac{-g_{1m}}{(g_{1DS} + g_{3DS} + g_{5DS})} .$$

Furthermore, the common-mode rejection ratio (CMRR) can be written as:

$$CMRR \equiv \left| \frac{A_{DM}}{A_{CM}} \right|$$
$$= \frac{g \cdot g_{1DS} + (g + 2g_{1m} + 2g_{1DS}) \cdot [g_{3L} + (g_{5m} + g_{5DS})]}{g \cdot [g_{1DS} + g_{3L} - (g_{5m} - g_{5DS})]} \tag{3.83}$$
$$\cong \frac{g \cdot g_{1DS} + 2g_{1m} \cdot 2g_{3m}}{g \cdot (g_{1DS} + g_{3DS} + g_{5DS})} \cong \frac{4g_{1m} \cdot g_{3m}}{g \cdot (g_{1DS} + g_{3DS} + g_{5DS})} .$$

Comparing to the CMRR without the cross-coupled transistors M5 and M6 [3]

$$CMRR \cong \frac{2g_{1m}}{g} , \tag{3.84}$$

the increase of CMRR by the factor $\frac{2g_{3m}}{g_{1DS} + g_{3DS} + g_{5DS}}$ is enormous because g_m is normally much larger than g_{DS}.

Unfortunately, Eq. (3.82) is only valid if both the transconductances of the devices M3 and M5 match exactly, which cannot be guaranteed in practice. In case that mismatch occurs between the transistors M3 and M5, the term $g_{3m} - g_{5m}$ from Eq. (3.82) can be either positive or negative. Moreover, if this term is negative and its value is larger than $g_{1DS} + g_{3DS} + g_{5DS}$, this leads to the differential-mode voltage gain A_{DM} to be positive. A proper operation of this circuit in this case is not possible. Other disadvantage of this kind of amplification is that the transient performance is too slow.

3.3.5 Principles and Design Techniques for Interpolating and Averaging

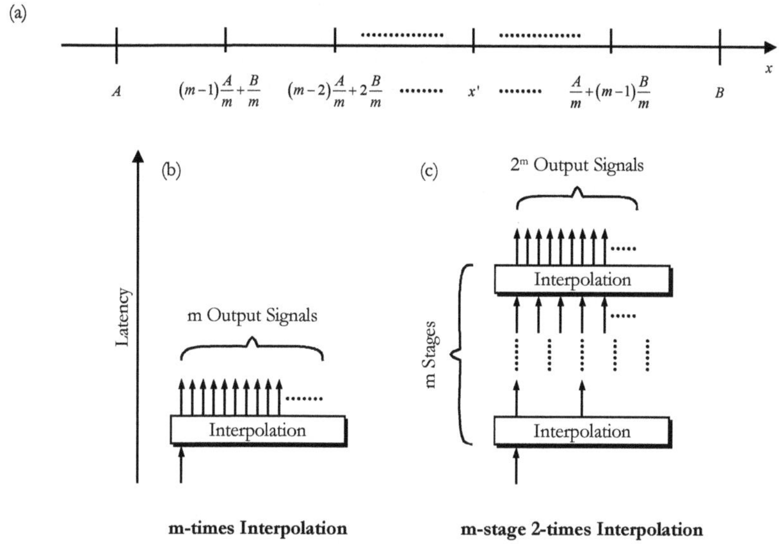

Fig. 3.19: Definitions of linear interpolation, the m-times and the m-stage interpolation

The *interpolation* [7] is a method to generate additional signals from the input signal of the interpolation stage. That is to say, the total number of usable signals can be increased by this method. Providing that the *interpolation error*, which is defined as the difference between the desired interpolated value and the actual value of a certain signal, can be well controlled, the resolution of the converter can be increased at low cost. Consequently, the design technique of interpolation has essential influence on reducing the power dissipation and the total die area of the complete converter. Since the methods of interpolation can be very different, the shrinking and complexity of the converter depends on the interpolating method that is applied. Generally speaking, the total number of usable signals is multiplied with the *interpolation factor* F_{ITPL} after the interpolation stage. Thus, the interpolator should be inserted close to the rear end of the converter, instead of placing close it to the front end, to ensure the efficiency of interpolation.

Depending on the electrical signal employed in the interpolation, methods of interpolation can be divided into two categories, namely, *voltage interpolation* [40, 71, 72] and *current interpolation* [7, 74, 75]. The advantage of voltage interpolation is its simplicity of realization. On the other hand, the current signal can yield more dynamic range than the voltage signal.

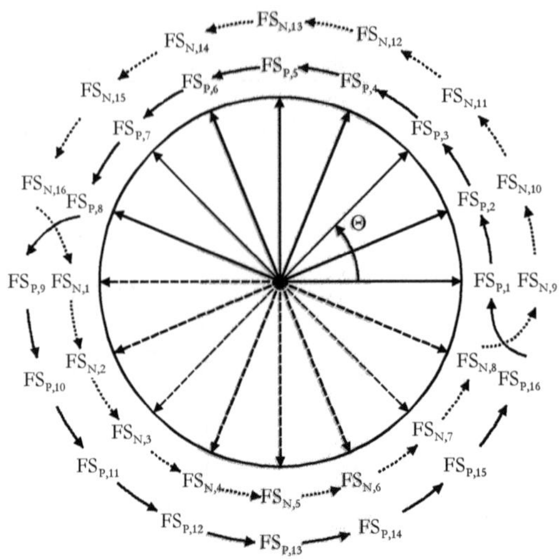

Fig. 3.20: Folding signals displayed in polar coordinate system

From the viewpoint of implementation, the voltage interpolation can be realized using a resistor string, like a voltage divider. Such kind of resistor will be called *interpolating resistor* R_{ITPL} in the later text. High-ohmic interpolating resistors can ensure the needed precision of the interpolated signals. But, this is contrary to the need of *averaging technique*. Sizing of the interpolating resistor will be treated together with the averaging technique.

In contrast to the voltage interpolation, the current interpolation generates the interpolated signals by using current mirroring. This kind of interpolating technique is complicated to implement, but, with the current trend of low-voltage design towards 2.5 V or even lower, its high dynamic range is becoming more and more important.

Besides the character of the input signal of interpolation, there is another criterion to distinguish two different types of interpolation. They are *multi-times interpolation* and *multi-stage interpolation*. The multi-stage interpolation is intrinsically a cascaded interpolation. Both the multi-times and the multi-stage interpolation treated in this book are kinds of *linear interpolation* (Fig. 3.19(a)).

Arithmetically the linear interpolation of the signal x in the range between two known values A and B can be observed as generating fraction amount of the single values A/m and B/m and using these fractions for linear approximation of the signal x for other values in this range. The value x' in Fig. 3.19(a) can be written as [7, 8]:

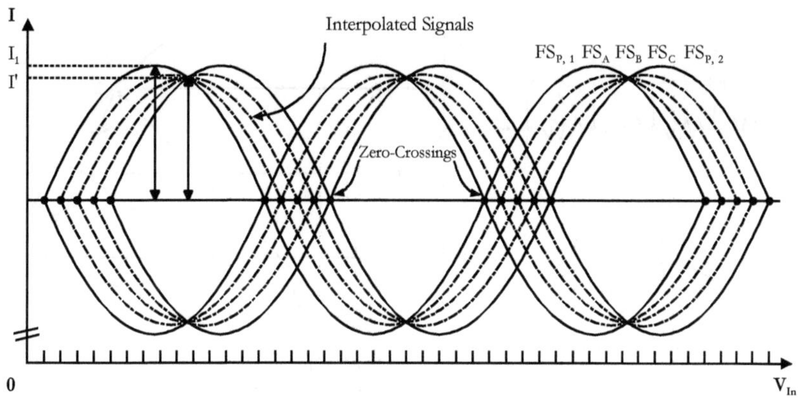

Fig. 3.21: Linear 4-times interpolation in the Cartesian coordinate system

$$x' = p \cdot \frac{A}{m} + q \cdot \frac{B}{m} , \tag{3.85}$$

where $0 \leq p, q \leq m$, $p + q = m$, and $p, q, m \in \mathbb{N}$.

The value x' represents the interpolated value. Thus, the interpolation error can be defined as the difference between the actual value x and the desired interpolated value x'.

The multi-times interpolation implies that the input signal is interpolated in order to obtain several output signals directly, as illustrated in Fig. 3.19(b). Extra *m-1* signals are generated by the input signals. This kind of interpolation is more efficient. But if *m* is much greater than 2, the risk of introducing an interpolation error is larger than for the 2-times interpolation. Details of calculation of the interpolation error will be discussed later.

On the other hand, the multi-stage interpolation has disadvantage of longer signal latency as it can be seen in Fig. 3.19(c). The interpolation error can be minimized, if only 2-time interpolation for every stage is used. Of course, it is possible that arrangements of multi-times interpolation and multi-stage interpolation are designed to be combined together. In this case, trade-offs have to be made between the efficiency of the interpolation and the potential interpolation error.

In order to describe the process of interpolation and its corresponding interpolation error mathematically, let us look at first the folding signals *FS* in Fig. 3.13. As can be seen in this figure, the folding signal $FS_{N,j}$ has the same position as the *virtual folding signal* $FS_{P,j+4}$. For the sake of simplifying discussion without loss of the continuity, we assume that the folding signal *FS*

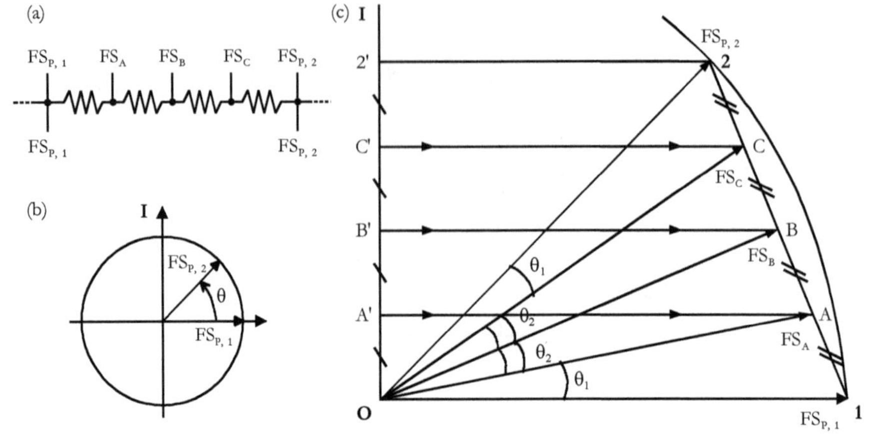

Fig. 3.22: Linear 4-times interpolation in the polar coordinate system

in Fig. 3.13 can be treated as sine-wave signal. Then the *x*-axis that carries V_{In} can be converted into polar angle Θ, which is usually used as a time-based variable and can be described by the *polar coordinate system*. Hence, the folding signals FS like in Fig. 3.13 can be displayed using the polar coordinate system as depicted in Fig. 3.20, supposing that there are totally only 8 folding signals $FS_1 \sim FS_8$.

Some general characteristics of the folding signals FS can be better observed in Fig. 3.20. The original *positive folding signals* $FS_{P,1} \sim FS_{P,8}$ are marked by arrows with solid lines outside the circle; on the other side, the original *negative folding signals* $FS_{N,1} \sim FS_{N,8}$ are marked by arrows with dotted lines outside the circle. These 16 folding signals $FS_{P,1} \sim FS_{P,8}$ and $FS_{N,1} \sim FS_{N,8}$ are spanned equidistantly over the whole circle and create the internal ring. In contrast to this, since the $FS_{N,1}$ can be also observed as $FS_{P,9}$, as discussed above, the folding signals $FS_{P,9} \sim FS_{P,16}$ and $FS_{N,9} \sim FS_{N,16}$ form the external ring.

In Fig. 3.20 all positive folding signals and their corresponding negative folding signals have 180° *"phase difference"*. The so-called phase difference in the polar coordinate system actually corresponds to a DC-offset in the *Cartesian coordinate system*. Second, if the folding signals are equidistantly positioned along the *x*-axis (Fig. 3.13), they are spaced equally at the same angle around the ring in Fig. 3.20. Hence, interpolation along the *x*-axis in Fig. 3.13

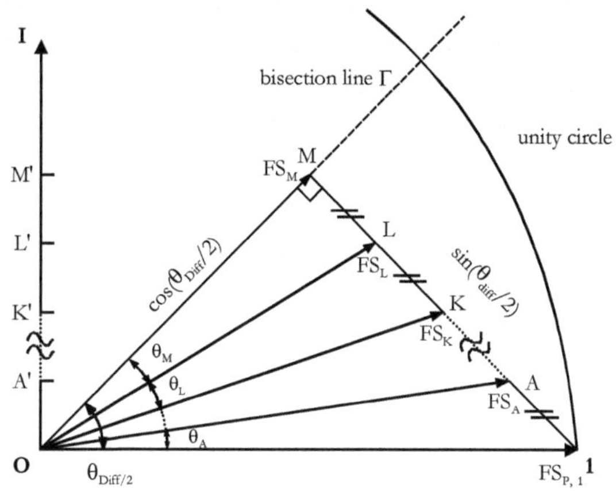

Fig. 3.23: Modeling for estimating Interpolation error

is equivalent to interpolation around the polar angle in Fig. 3.20. The model shown in Fig. 3.20 can be used to investigate the interpolation error.

Fig. 3.21 shows two folding signal $FS_{P,\,1}$, $FS_{P,\,2}$ and the corresponding *interpolated signals*, marked with FS_A, FS_B, and FS_C. For the comparator, only the positioning of *zero-crossings* is of importance. The amplitude difference between I_1 and I' is not a crucial problem.

The creation of this amplitude difference $(I_1 - I')$ can be better understood considering in the polar coordinate system, as illustrated in Fig. 3.22. This figure shows a simple model of 4-times voltage interpolation (Fig. 3.22(a)); the input signals are $FS_{P,\,1}$ and $FS_{P,\,2}$ and three interpolated signals are FS_A, FS_B, and FS_C. Since the input signals $FS_{P,\,1}$ and $FS_{P,\,2}$ can also be used as output signals and every input signal can be used twice, the interpolation degree can be considered as 4-times interpolation, although only three new signals have been generated by the interpolation.

Fig. 3.22(b) shows the input signals $FS_{P,\,1}$ and $FS_{P,\,2}$ in the polar coordinate system. The three interpolated signals FS_A, FS_B, and FS_C are illustrated in Fig. 3.22(c). The generation of interpolated signals is described in the following.

According to Eq. (3.86), the interpolation is addition of the fractions of the input signals. Hence, only the amplitude of the input signals (here: folding signals FS) has to be considered.

If the folding signals can be described as sine-wave functions as shown in Fig. 3.21 (Cartesian coordinate system) and Fig. 3.22(c) (polar coordinate system), only the components of the I-axis have influence on the interpolation process. Hence, the signals $FS_{P,1}$ and $FS_{P,2}$ must be projected onto the I-axis as illustrated in Fig. 3.22(c), which corresponds to the cosine component in the polar coordinate system. The total difference between signals $FS_{P,1}$ and $FS_{P,2}$ is $\overline{O2'}$, which is divided into four equal parts, $\overline{OA'}$, $\overline{A'B'}$, $\overline{B'C'}$, and $\overline{C'2'}$. The projection onto the line $\overline{12}$ by these four sections defines the real values of the interpolated signals.

Since $\theta_2 > \theta_1$, the interpolated signals FS_A and FS_C are expected to have an interpolation error. To the contrary, the interpolated signal FS_B has no interpolation error because the interpolated signal FS_B is onto the bisection line of the angle $\angle(FS_{P,1}, FS_{P,2})$ and the desired value is equal to the actual value at this point.

Henceforth, consider the general case of interpolation error. For an $m+n$-bit folding and interpolating A/D converter with the folding factor $F_F = 2^m$, the LSB can be described under the polar coordinate system as:

$$LSB = \frac{180°}{2^n} . \tag{3.86}$$

Using the same definitions as in Chapter 3.3.1, the phase difference θ_{Diff} between two neighboring folding signals, which are the output signals of the folding stage, can be written as:

$$Phase\ Difference\ \theta_{Diff} = LSB \times 2^{n_2} = \frac{180°}{2^n} \times 2^{n_2} = \frac{180°}{2^{n_1}} , \tag{3.87}$$

where n_2 is the *interpolation resolution.*

Since the interpolation is symmetrical to the bisecting line of the angle $\angle(FS_{P,1}, FS_{P,2})$, only the difference between the folding signal $FS_{P,1}$ and the bisecting line Γ have to be taken into account.

Furthermore, for a given amplitude, i.e. the folding signals $FS_{P,1}$ and $FS_{P,2}$ lie on an unity circle, the lengths $|\overline{OM}|$ and $|\overline{1M}|$ can be written as:

$$|\overline{OM}| = \cos\left(\frac{\theta_{Diff}}{2}\right) , \tag{3.88}$$

and

$$\left|\overline{1M}\right| = \sin\left(\frac{\theta_{Diff}}{2}\right). \tag{3.89}$$

Hence, the angles θ_A, ..., θ_L, and θ_M in Fig. 3.23 can be described as:

$$\begin{aligned}\theta_M &= \tan^{-1}\left(\frac{\left|\overline{LM}\right|}{\left|\overline{OM}\right|}\right) = \tan^{-1}\left[\frac{\sin\left(\frac{\theta_{Diff}}{2}\right)}{\left(\frac{2^{n_2}}{2}\right)\cdot\cos\left(\frac{\theta_{Diff}}{2}\right)}\right] \\ &= \tan^{-1}\left[\frac{\tan\left(\frac{\theta_{Diff}}{2}\right)}{2^{n_2-1}}\right],\end{aligned} \tag{3.90}$$

$$\begin{aligned}\theta_L &= \tan^{-1}\left(\frac{\left|\overline{KM}\right|}{\left|\overline{OM}\right|}\right) - \theta_M \\ &= \tan^{-1}\left[2\cdot\left[\frac{\sin\left(\frac{\theta_{Diff}}{2}\right)}{\left(\frac{2^{n_2}}{2}\right)\cdot\cos\left(\frac{\theta_{Diff}}{2}\right)}\right]\right] - \tan^{-1}\left[1\cdot\left[\frac{\sin\left(\frac{\theta_{Diff}}{2}\right)}{\left(\frac{2^{n_2}}{2}\right)\cdot\cos\left(\frac{\theta_{Diff}}{2}\right)}\right]\right] \\ &= \tan^{-1}\left[\frac{\tan\left(\frac{\theta_{Diff}}{2}\right)}{2^{n_2-2}}\right] - \tan^{-1}\left[\frac{\tan\left(\frac{\theta_{Diff}}{2}\right)}{2^{n_2-1}}\right],\end{aligned} \tag{3.91}$$

and

$$\begin{aligned}\theta_K &= \tan^{-1}\left(\frac{\left|\overline{JM}\right|}{\left|\overline{OM}\right|}\right) - (\theta_L + \theta_M) \\ &= \tan^{-1}\left[3\cdot\left[\frac{\sin\left(\frac{\theta_{Diff}}{2}\right)}{\frac{2^{n_2}}{2}\cdot\cos\left(\frac{\theta_{Diff}}{2}\right)}\right]\right] - \tan^{-1}\left[2\cdot\left[\frac{\sin\left(\frac{\theta_{Diff}}{2}\right)}{\frac{2^{n_2}}{2}\cdot\cos\left(\frac{\theta_{Diff}}{2}\right)}\right]\right] \\ &= \tan^{-1}\left[3\cdot\frac{\tan\left(\frac{\theta_{Diff}}{2}\right)}{2^{n_2-1}}\right] - \tan^{-1}\left[\frac{\tan\left(\frac{\theta_{Diff}}{2}\right)}{2^{n_2-2}}\right],\end{aligned} \tag{3.92}$$

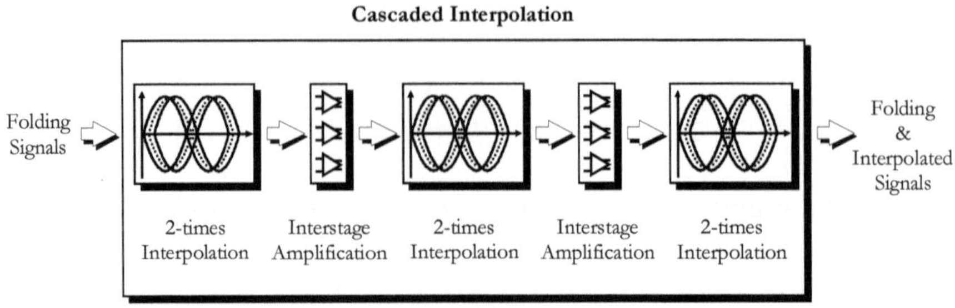

Fig. 3.24: Cascaded 3-stage, 2-times interpolation stage

and so forth.

The DNL at point M, according to Eq. (2.20), is then

$$DNL_M \equiv \frac{|LSB| - |(code\ width)_M|}{|LSB|} = \frac{\left(\frac{180°}{2^n} - \theta_M\right)}{\frac{180°}{2^n}}$$
$$= 1 - \frac{2^n}{180°} \cdot \tan^{-1}\left[\frac{\tan\left(\frac{\theta_{Diff}}{2}\right)}{2^{n_2 - 1}}\right]. \tag{3.93}$$

The INL can be obtained by using Eq. (2.21).

Furthermore, supposing that only *multi-stage 2-times interpolation,* instead of *single stage, multi-times interpolation* (see Fig. 3.19(b) and Fig. 3.19(c)), are used in the interpolation stage, the interpolation error, based on the viewpoint of DC analysis, will be minimized. The amplitude of these interpolated signals can be controlled by using *interstage amplifiers*. Fig. 3.24 depicts an interpolation scheme based on a cascaded 3-stage 2-times interpolation stage, which is going to be implemented in this book.

Averaging technique was introduced by Mr. Kattman in 1991 [91] and improved by Mr. Bult [71-73]. The basic principle of averaging technique is to "de-sharpen" the effects of the mismatch and the offset from the amplifiers by adding resistors to connect the output signals of the interstage amplifiers (see Fig. 3.24). An example of inserting the *averaging resistor* R_A between the neighboring amplifiers inside the interstage amplification is depicted in Fig. 3.25.

The load in Fig. 3.25 can be either a passive load, e.g. a resistor, or an active load, e.g. a current source. The differences, the assets, and drawbacks,

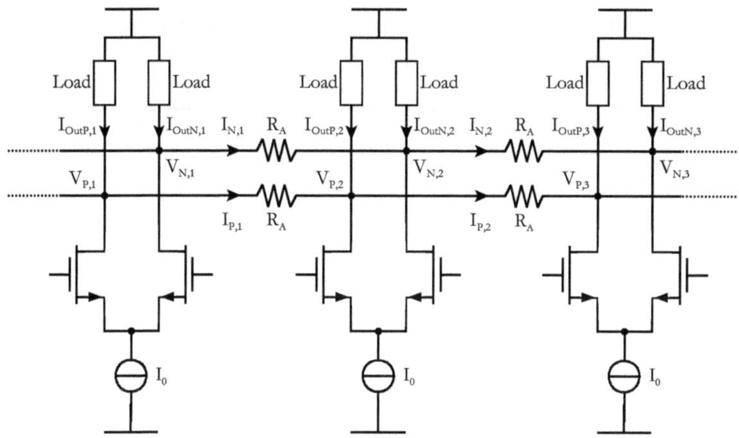

Fig. 3.25: Averaging circuits

presented by Mr. Kattman in 1991 [91] and Mr. Bult in 1998 [73], of using a physical resistor or using current source as the load of the amplifier are summarized in Table 3.2. Investigations made by Mr. Bult [71-73] showed that setting the resistors R_L and R_A to be equal is a good trade-off between the output impedance of the amplifier and the averaging effect. We shall now consider the averaging more closely and try to determine optimum values for R_L and R_A.

Let us assume a simple model for averaging shown in Fig. 3.26. That is, we assume that only three amplifiers (folding amplifier or interstage amplifier) operate in linear or quasilinear region, the rest is saturated (i.e. the output is "clipped"). Then we can calculate the output voltage at the node k as:

$$V_{Out,k} = -\frac{I_{Out,k} + \alpha \cdot I_{Out,k-1} + \alpha \cdot I_{Out,k+1}}{1+2\alpha} \cdot R_L \,, \tag{3.94}$$

where $\alpha = R_L/(R_A + R_L)$. Note that we have neglected the power supply voltages, since we are interested only in small-signal operation. Assuming that $I_{Out,j} = G_m \cdot V_{i,j}$ for j=k-1, k, and k+1 we can write:

$$V_{Out,k} = -\frac{V_{i,k} + \alpha \cdot V_{i,k-1} + \alpha \cdot V_{i,k+1}}{1+2\alpha} \cdot G_m R_L \,, \tag{3.95}$$

where G_m is the transconductance of the amplifier employed and $V_{i,k}$ the input voltage of the kth amplifier connected to the node k.

Table 3.2: Assets and drawbacks of physical resistors employed as passive load and current sources as active load

	Assets	Drawbacks
Physical resistor R_L (with smaller R_A)	- Amplifier offsets ↓ - Better averaging effect	- Output impedance of amplifier ↓ - DC voltage gain ↓ - Comparator offsets ↑ (connected to $V_{P,x}$ and $V_{N,x}$ in Fig. 3.25)
Trade-off	Set R_L and R_A to be equal	
Current source (output resistance r_{DS})	- RL replaced by r_{DS} - Gain is more determined by $I_{P,x} \cdot R_A$ and $I_{N,x} \cdot R_A$ - Gain and averaging decoupled to first order extent	- Nonlinear relationship between r_{DS} and I_{Out} leads to nonlinear averaging on $V_{P,x}$ and $V_{N,x}$
Approach	Lower R_A until the right gain is reached	

Based on this result let us estimate the DNL for the case that the kth amplifier exhibits an error voltage $V_{Error,k}$ and is connected to a reference voltage $V_{Ref,k}$, where $V_{Ref,k}$ can be calculated as $V_{Ref,k} = (1-k/N) \cdot V_{Ref}$. First we obtain (see Eq. (2.20))

$$\left|\Delta V_{Out,k}\right| = \left|V_{Out,k} - V_{Out,k-1}\right| = \frac{\begin{bmatrix} \alpha \cdot V_{i,k+1} + (1-\alpha) \cdot V_{i,k} - \\ (1-\alpha) \cdot V_{i,k-1} - \alpha \cdot V_{i,k-2} \end{bmatrix}}{1+2\alpha} \cdot G_m R_L \,. \tag{3.96}$$

Then we insert $V_{i,j} = V_{In} - V_{Ref,k} + V_{Error,k}$, which yields

$$\left|\Delta V_{Out,k}\right| = \begin{bmatrix} \alpha \cdot V_{Error,k+1} + (1-\alpha) \cdot V_{Error,k} - \\ \dfrac{(1-\alpha) \cdot V_{Error,k-1} - \alpha \cdot V_{Error,k-2}}{1+2\alpha} + \dfrac{V_{Ref}}{N} \end{bmatrix} \cdot G_m R_L \,, \tag{3.97}$$

for $V_{Ref,k} = (1-k/N) \cdot V_{Ref}$.

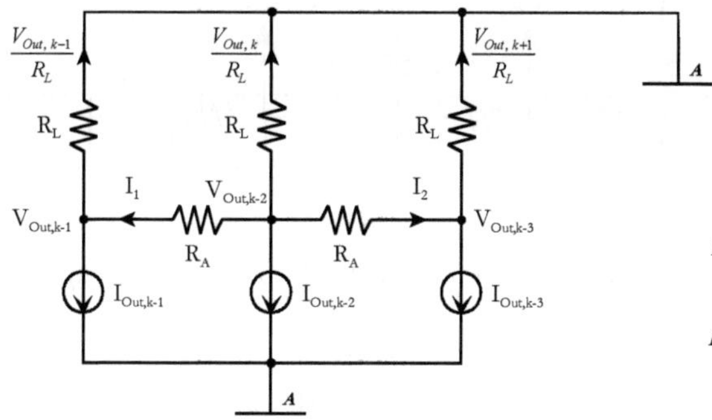

$$I_1 = \frac{V_{Out,\,k} - V_{Out,\,k-1}}{R_A}$$

$$I_2 = \frac{V_{Out,\,k} - V_{Out,\,k+1}}{R_A}$$

Fig. 3.26: Model for averaging technique

If we further assume that all amplifiers have a constant offset ΔV_{OS}, but the kth amplifier exhibits an offset mismatch ΔV_{OS} (i.e. $V_{Error,j} = V_{OS}$ for $j \neq k$, $V_{Error,\,k} = V_{OS} + \Delta V_{OS}$), we obtain

$$\left|\Delta V_{Out,\,k}\right| - \left(\frac{1-\alpha}{1+2\alpha}\cdot\Delta V_{OS} + \frac{V_{\mathrm{Re}f}}{N}\right)\cdot G_m R_L \ , \tag{3.98}$$

which clearly proves that the averaging reduces the effect of the device mismatch on the DNL. This can be seen when considering the case of $R_A = \infty$ (i.e. $\alpha = 0$), i.e.

$$\left|\Delta V_{Out,\,k}\right|_{\alpha=0} = +\left(\Delta V_{OS} + \frac{V_{\mathrm{Re}f}}{N}\right)\cdot G_m R_L \ , \tag{3.99}$$

that is when no averaging is performed. Nevertheless, the price paid for this improvement is the degradation of the INL. If we determine the INL according to Eq. (2.21), namely

$$\left|\sum_{j=k-1}^{k+1} \Delta V_{Out,j}\right| = \left(\frac{\alpha\cdot\Delta V_{OS}}{1+2\alpha} + \frac{3V_{Ref}}{N}\right)\cdot G_m R_L \ . \tag{3.100}$$

We can see that α affects adversely the INL (i.e. the INL is better without averaging). This is due to the fact that with rising α the error "leaks" to neighboring amplifiers, although its peak decreases. Optimization is difficult if we do not know, what is more important for a particular application, i.e.

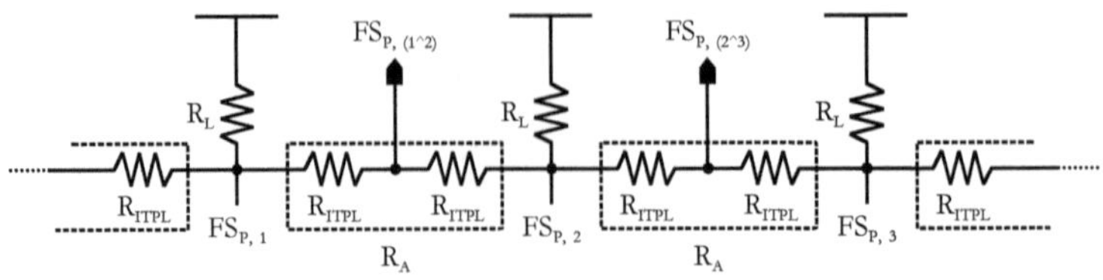

Fig. 3.27: Principle of mixture of interpolation and averaging effects

DNL or INL, but minimizing the root-mean-square sum of both, DNL and INL, yields $\alpha_{Opt} = 3/4$.

We can now estimate also the DNL improvement due to averaging for a statistical device mismatch of all amplifiers expressed as $\overline{\Delta V_{OS}^2}$ using the above given equations, namely, as variance

$$\overline{\Delta V_{Out,\,k}^2} = 2\frac{\alpha^2 + (1-\alpha)^2}{(1+2\alpha)^2} \cdot \overline{\Delta V_{OS}^2} \,. \tag{3.101}$$

The improvement can be calculated considering $\alpha = 3/4$ and $\alpha = 0$ (i.e. no averaging), which yields an improvement of factor of 10 referred to noise power, i.e. 10 dB or 1.66 bit.

In this book, the approach is to combine the functions of interpolation and averaging together. However, this approach requires use of *high-ohmic polysilicon resistors* in the CMOS process employed for realization (instead of *low-ohmic polysilicon* used for transistor gates). The principle of combining the 2-times interpolation with averaging effect can be seen in Fig. 3.27. The resistors R_L and R_A in Fig. 3.27 are designed to be equal to meet the demands of the output impedance of the amplifier and the averaging effect. This circuit has been used in cascaded interpolation stage of the implemented A/D converter, as shown in Fig. 3.24.

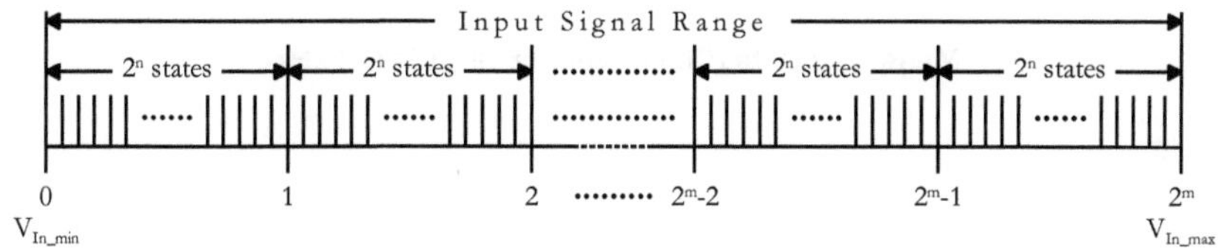

Fig. 3.28: Distribution of reference voltages for coarse and fine converters

3.4 Subsystem: Coarse Converter

3.4.1 Introduction

The coarse converter of the folding and interpolating A/D converter divides the input signal range into several intervals and provides the heading significant bits of the complete converter. In addition to the coarse converter, the fine converter resolves each interval separately and supplies the remaining bits for the outputs of the entire converter.

An alternative function of the coarse converter is to detect the out-of-range input signal, including both the overflow and underflow situations. Hence, as an example of m-bit coarse converter, the total number of the comparators inside the coarse converter is $(2^m + 1)$. The distribution of the reference voltages used in the m-bit coarse converter and n-bit fine converter is shown in Fig. 3.28. The input signal range is divided into $(2^m \cdot 2^n)$ intervals.

Supposing that an interpolation technique with the interpolating factor $F_{ITPL} = 2^{n_2}$ is applied, then the number of needed reference voltages is only $(2^m \cdot 2^{n-n_2})$, as discussed in the section on fine converter stage in this chapter.

3.4.2 Coarse Converter Topology

The coarse converter can be just a flash A/D converter if the coarse resolution m is not too high. Assuming that the folding and interpolating A/D converter is designed with both sample-and-hold amplifier and synchronization stage, then an extra delay stage to align the coarse converter with the fine converter is not necessary. But, if the folding and interpolating A/D converter is designed without the sample-and-hold amplifier as its analog front-

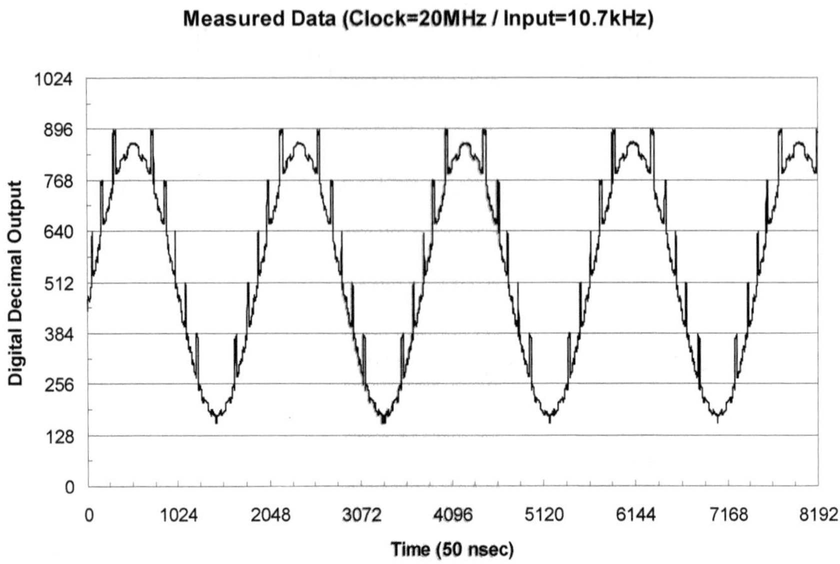

Fig. 3.29: Folding and interpolating A/D converter with only THA as synchronization

end, a delay stage between the input signal and the coarse comparators to prevent the signal racing between the two converters is a suitable arrangement [92]. Otherwise, the latency caused by the coarse converter will be much shorter than that by the fine converter. The output signal might not be well synchronized.

The sample-and-hold or, better to say, track-and-hold amplifier acquires and holds the input signal and it also works as a "pace maker" for the following circuits. Thus, the THA can be considered as the primary synchronization stage. But the acquired signal itself has glitches when the clock signal switches, which might cause signal racing in the following stages. Besides this, the pedestal error and the droop rate of the THA can deteriorate the performance of the converter, especially when the acquired signal is near the interval edges. As an example, a measured result of such synchronization error occurring close to the interval edges is given in Fig. 3.29. The measured data shows that the converter with THA as the only synchronization stage exhibits synchronization errors at the interval edges. Thus, circuits for the synchronization cannot be avoided, although the THA already provides the synchronization function.

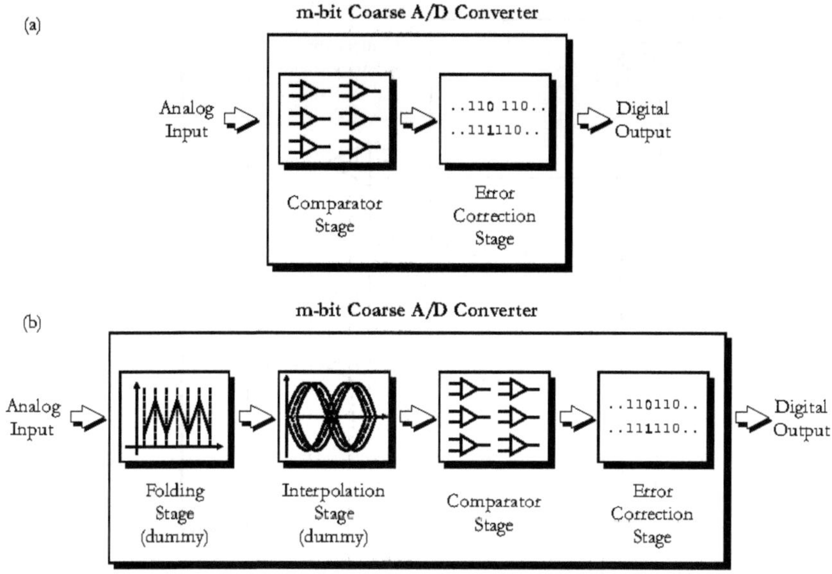

Fig. 3.30: Possible structures of the coarse converter

In addition, using the THA can acquire and "freeze" the input signal at a certain signal level and the problem of signal racing in both, coarse and fine, converters is diminished. Moreover, the increased internal frequency inside the folder stage is limited only when the clock signal switches. Thus what is now more important in the converter design is balancing the latencies of both converters rather than minding this increased internal frequency. Fig. 3.30 shows two possible structures of the coarse converter. The first one (Fig. 3.30(a)) can be operated with sampled or unsampled input signal. In this case, the coarse converter itself is designed like a flash A/D converter without extra signal delay circuits, but in the case of unsampled signal, more care must be executed in the system design than for systems with sampled input signal.

In the case of a conversion system without the sample-and-hold amplifier as an analog front-end, it is better to insert an analog delay circuit in front of the comparator stage to compensate for the different latencies between the coarse and the fine converter. One example suggested by Mr. Wenzek [92] is to insert cascaded dummy differential amplifiers with similar structure like those in the folding stage of the fine converter. An example of the block diagram is depicted in Fig. 3.30(b). The advantage of such a structure is that it avoids the use of high-speed THA, which is the most critical component in high-speed conversion. Besides, the power dissipation is also reduced in this

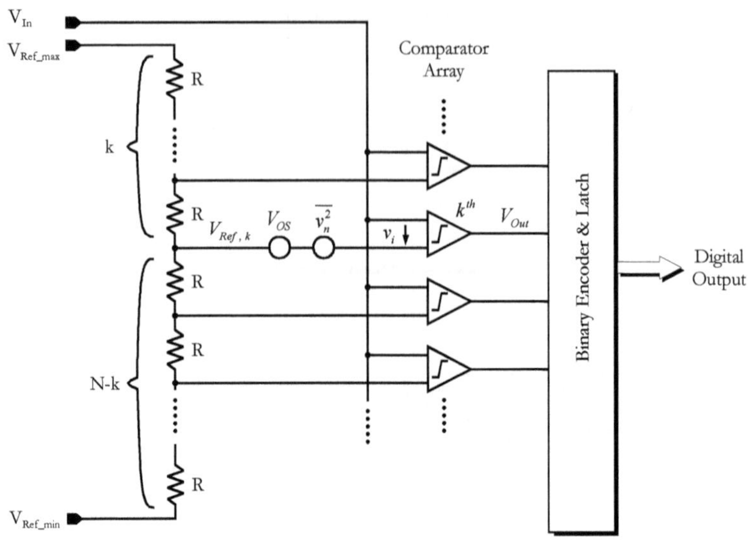

Fig. 3.31: Flash A/D converter

case because the THA stage consumes more power than these two dummy stages. The balance sheet for the power dissipation can be found in the section for power management in Chapter 4.

The designs for the comparator stage and the error correction stage are essentially similar to those in the fine converter and will be treated in the following sections.

3.5 Comparator Stage

The comparator stage forms an essential part of the flash A/D converter used here as a coarse converter and, therefore, deserves a thorough analysis.

3.5.1 Comparator Analysis

Let us consider a simple flash A/D converter block diagram illustrated in Fig. 3.31. It contains a resistor string to generate a set of reference voltages, a comparator array, and logic to encode the comparator signals. This figure also contains error sources that determine the accuracy of this type of A/D converter: *time-independent errors* are summed up in the source V_{OS}, while all *time-dependent error* sources are included in the *random noise source* $\overline{v_n^2}$. In

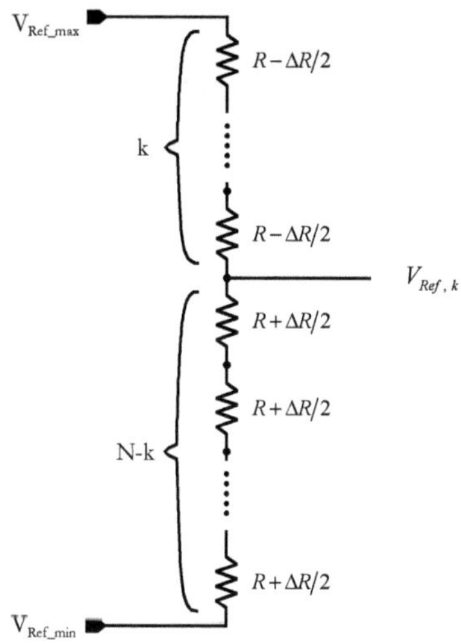

Fig. 3.32: Resistor string with "worst case" mismatch

circuit design terminology the source V_{OS} can be interpreted as comparator input offset voltage plus reference offset voltage, while the source $\overline{v_n^2}$ includes comparator input noise and noise of the reference voltage.

For an n-bit resolution this converter requires N resistors and $(N-1)$ comparators, where $N=2^n$. For the kth comparator, the input voltage can be described as:

$$v_i = V_{In} - V_{Ref,k} + V_{Error}\,, \tag{3.102}$$

where

$$V_{Ref,k} = \left(1-\frac{k}{N}\right)\cdot\left(V_{Ref\_max} - V_{Ref\_min}\right) = \left(1-\frac{k}{N}\right)\cdot V_{Ref} \tag{3.103}$$

and

$$V_{Ref} \equiv V_{Ref\_max} - V_{Ref\_min}\,. \tag{3.104}$$

Note that if all error sources V_{OS} were equal for all comparators, the resolution of the A/D converter would not suffer. Only statistical mismatch ΔV_{OS} degrades the resolution. Hence:

$$V_{Error}^2 = \Delta V_{OS}^2 + \overline{v_n^2} \; . \tag{3.105}$$

The comparator switches its logic output to high or low for $v_i > 0$ or $v_i < 0$, respectively. To meet the resolution requirements on differential nonlinearity, the error voltage must be smaller than half the LSB value, hence:

$$V_{Error,\ \max} \leq \frac{V_{Ref}}{2 \cdot N} = \frac{V_{Ref}}{2^{n+1}} = \frac{LSB}{2} \; . \tag{3.106}$$

As stated above, the error source V_{Error} thus plays an important role in determining the converter accuracy. As stated above, the time-independent error source ΔV_{OS} includes errors caused by the resistor string mismatch $\Delta V_{OS,\ Ref}$ and offset mismatch voltage of the comparator $\Delta V_{OS,\ CMP}$. We can identify the worst case for the reference voltage of the kth comparator. If, e.g., all k resistors exhibit a deviation of $-\Delta R/2$ and all remaining $(N-k)$ resistors have a deviation of $+\Delta R/2$ (see Fig. 3.32). Then we can write $\Delta V_{OS,\ Ref,\ \max}$ as:

$$\begin{aligned} \Delta V_{OS,\ Ref,\max} &= \left| V_{Ref,\ k,\ \max} - V_{Ref,\ k} \right| \\ &= \left| \frac{(N-k)\cdot\left(R + \frac{\Delta R}{2}\right)}{(N-k)\cdot\left(R+\frac{\Delta R}{2}\right) + k\cdot\left(R - \frac{\Delta R}{2}\right)} - \frac{N-k}{N} \right| \cdot V_{Ref} \\ &= \left| \frac{N-k}{N} \cdot \frac{k \cdot \frac{\Delta R}{R}}{N + (N-2k)\cdot \frac{\Delta R}{2R}} \right| \cdot V_{Ref} \; . \end{aligned} \tag{3.107}$$

The maximum value can be estimated for $k = N/2$ to be $\Delta V_{OS,\ Ref,\ \max} = \frac{V_{Ref}}{4} \cdot \left| \frac{\Delta R}{R} \right|$, and it is at least partially technology-dependent while $\Delta V_{OS,\ CMP,\ \max}$ is solely technology-dependent. It must be noted, however, that in practice the main contribution to the resistor mismatch appears to be the contact resistance variations rather than geometry variations.

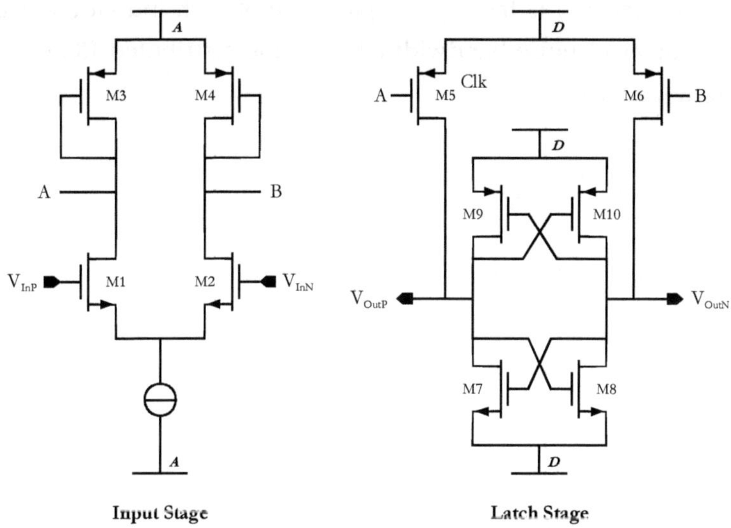

Fig. 3.33: Schematic of a high-speed comparator

Determination of noise contribution $\overline{v_n^2}$ is, of course, a much tougher task because a comparator is essentially a nonlinear circuit. We can help it by linearization of the circuit just for the initial time period, i.e. when the comparator is making its decision. This is especially true for the comparator considered here which consists of a preamplifier and a latch stage. The noise of the resistor string, on the other hand, is somewhat easier to be estimated.

Considering the fact that each resistor R shown in Fig. 3.32 exhibits a *thermal noise* $\overline{v_{nR}^2} = 4kTR \cdot \Delta f$, where Δf is the *equivalent noise bandwidth,* we can quickly derive the maximum noise voltage, namely $\overline{v_{n,\,Ref,\,\max}^2} = N \cdot \overline{v_{nR}^2}$. The determination of Δf, however, is not easy because the resistor string represents a distributed RC line due to the distributed loading of the resistor by the input capacitance C_{In} of the comparator array (this assumption neglects any parasitic capacitance of the resistor string to the substrate).

For $N \gg 1$ we can use the formula for the transfer function known for distributed RC networks, i.e.

$$H(s) = \frac{1}{\cosh\sqrt{sRC_{In}}}, \tag{3.108}$$

where $s = j\omega$ and C_{In} is the input capacitance of each individual comparator. The equivalent noise bandwidth $B_{Eq.}$ of the distributed RC line can be then determined as:

$$B_{Eq.} = \frac{1}{2\pi}\int_0^{\infty} \frac{1}{\cosh^2 \sqrt{j\omega RC}}\, d\omega \ . \tag{3.109}$$

The evaluation of this integrals yields:

$$B_{Eq.} = \frac{\ln 2}{\pi RC_{In}} \ , \tag{3.110}$$

and, consequently

$$\overline{v^2_{n,\ \mathrm{Re}\, f,\ \max}} = N \cdot 4kTR \cdot \frac{\ln 2}{\pi RC_{In}} = \frac{4N \cdot \ln 2}{\pi} \cdot \frac{kT}{C_{In}} \ . \tag{3.111}$$

Let us now turn our attention to the comparator (Fig. 3.33). This comparator consists of differential input stage and a latch stage. While its large-signal behavior is somewhat complex to describe (this will be done in the next chapter), here we try to derive its small-signal properties in order to determine accuracy requirements in terms of offset, mismatch, and noise. Although essentially a nonlinear circuit, its instantaneous behavior for the moment of decision can be described using the small-signal transfer function as:

$$A_{V,\ tot}(s) = \frac{g_{m1}}{g_{m3} + sC_A} \cdot \frac{g_{m5} \cdot (g_{m7} + g_{m9} + sC_B)}{(sC_B)^2 - (g_{m7} + g_{m9})^2} \ , \tag{3.112}$$

assuming that the device pairs *M1* & *M2*, *M3* & *M4*, *M5* & *M6*, *M7* & *M8*, and *M9* & *M10* are matched. This makes it possible to express the output noise as:

$$\overline{v^2_{Out,\ n,\ tot}} = 2 \cdot \left\{ \begin{aligned} &\left[\overline{v^2_{n1}} \cdot \left(\frac{g_{m1}}{g_{m3}} \right)^2 + \overline{v^2_{n3}} + \overline{v^2_{n5}} \right] \cdot \\ &\frac{g^2_{m5}}{(g_{m7} + g_{m9})^2} + \overline{v^2_{n7}} + \overline{v^2_{n9}} \end{aligned} \right\} \ , \tag{3.113}$$

where

$$\frac{\overline{v_{ni}^2}}{\Delta f} = \frac{k'}{C_{OX} \cdot W_i \cdot L_i \cdot f} + 4kT \cdot \frac{2}{3g_{mi}}, \tag{3.114}$$

for transistor Mi. The equivalent noise bandwidth for the thermal noise of the preamplifier *M1~M4* is $g_{m3}/(4C_A)$, where C_A is the total capacitance at the node A (including the gate capacitance of all devices connected to this node) and for the latch it is $(g_{m7}+g_{m9})/(4C_{Out})$, where C_{Out} is the capacitance at the output node including the total gate capacitance and capacitive loading at this node. The output noise is then:

$$\overline{v_{Out,\,n,\,tot}^2} = 2 \cdot \left\{ \begin{array}{l} \left[\begin{array}{l} \frac{k'}{C_{OX}} \left(\frac{1}{W_1 L_1} \left(\frac{g_{m1}}{g_{m3}} \right)^2 + \frac{1}{W_3 L_3} \right) \ln \frac{g_{m3} t_P}{C_A} + \\ \frac{k'}{C_{OX} W_5 L_5} \ln \frac{(g_{m7}+g_{m9}) t_P}{C_{Out}} \\ + \frac{2kT}{3C_{Out}} \left(\frac{g_{m7}+g_{m9}}{g_{m5}} \right) + \frac{2kT}{3C_A} \left(\frac{g_{m1}+g_{m3}}{g_{m3}} \right) \end{array} \right] \cdot \\ \frac{g_{m5}^2}{(g_{m7}+g_{m9})^2} + \frac{2kT}{3C_{Out}} \left(2 + \frac{g_{m7}}{g_{m9}} + \frac{g_{m9}}{g_{m7}} \right) + \\ \frac{k'}{C_{OX}} \left(\frac{1}{W_7 L_7} + \frac{1}{W_9 L_9} \right) \ln \frac{(g_{m7}+g_{m9}) t_P}{C_{Out}} \end{array} \right\}, \tag{3.115}$$

where the *conversion time* t_P can be approximated for large C_{Out} using the step response based on the linearized small-signal comparator transfer function introduced above:

$$t_{P,\,\max} \approx \frac{C_{Out}}{g_{m7}+g_{m9}} \cdot \ln\left(1+2^n\right), \tag{3.116}$$

which can be for large n further simplified to:

$$t_{P,\,\max} \approx \frac{n \cdot C_{Out}}{g_{m7}+g_{m9}} \cdot \ln 2 \, . \tag{3.117}$$

Hence, we can write for the *equivalent input noise voltage* of the comparator:

$$\overline{v_{n,\,CMP}^2} = \frac{\overline{v_{Out,\,n,\,tot}^2}}{\left(\frac{g_{m1}}{g_{m3}} \right)^2 \cdot \left(\frac{g_{m5}}{g_{m7}+g_{m9}} \right)^2}, \tag{3.118}$$

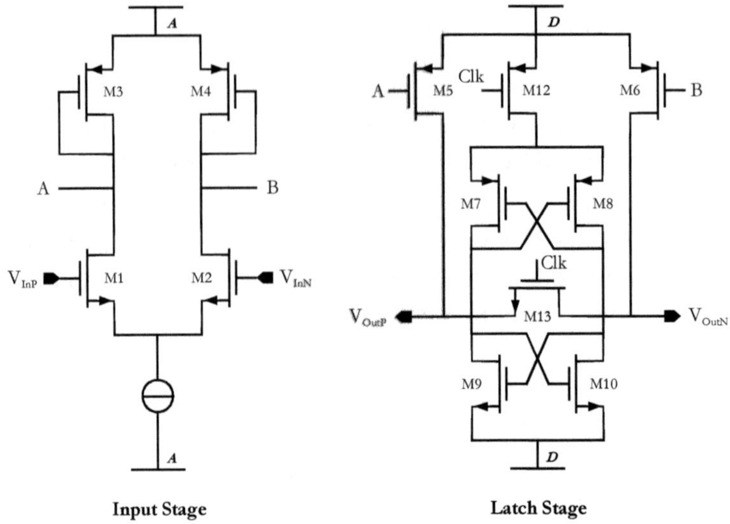

Fig. 3.34: Schematic of a comparator with clock

and further,

$$\overline{v_{n,\,CMP}^{2}}=2\cdot\left\{\begin{array}{l}\dfrac{k'}{C_{OX}}\left[\dfrac{1}{W_1L_1}+\dfrac{\left(g_{m3}/g_{m1}\right)^2}{W_3L_3}\right]\ln\left(\dfrac{\xi C_{Out}\cdot g_{m3}}{C_A\cdot\left(g_{m7}+g_{m9}\right)}\right)+\\ \dfrac{2kT}{3C_A}\left(1+\dfrac{g_{m3}}{g_{m1}}\right)\dfrac{g_{m3}}{g_{m1}}+\dfrac{k'}{C_{OX}W_5L_5}\left(\dfrac{g_{m3}}{g_{m1}}\right)^2\ln\xi+\\ \dfrac{k'}{C_{OX}}\left(\dfrac{1}{W_7L_7}+\dfrac{1}{W_9L_9}\right)\left(\dfrac{g_{m3}}{g_{m1}}\right)^2\left(\dfrac{g_{m7}+g_{m9}}{g_{m5}}\right)^2\ln\xi+\\ \dfrac{2kT}{3C_{Out}}\cdot\left(\dfrac{g_{m3}}{g_{m1}}\right)^2\left(\dfrac{g_{m7}+g_{m9}}{g_{m5}}\right)\\ \left[1+\left(2+\dfrac{g_{m7}}{g_{m9}}+\dfrac{g_{m9}}{g_{m7}}\right)\left(\dfrac{g_{m7}+g_{m9}}{g_{m5}}\right)\right]\end{array}\right\}, \tag{3.119}$$

where

$$\xi=n\cdot\ln 2\,. \tag{3.120}$$

Note that in order to minimize the noise and to maximize the DC voltage gain it is required $g_{m1}\gg g_{m3}$, $g_{m5}\gg g_{m7}+g_{m9}\gg g_{m3}$, and $g_{m7}=g_{m9}$. Since for the noise it is valid

$$\overline{v_n^2} = \overline{v_{n,\,\mathrm{Ref},\,\max}^2} + \overline{v_{n,\,CMP}^2}\,, \tag{3.121}$$

we can write the resolution requirement for the flash A/D converter as:

$$V_{Error,\,\max}^2 = \left\{ \begin{array}{l} \Delta V_{OS,\,\mathrm{Ref},\,\max}^2 + \Delta V_{OS,\,CMP,\,\max}^2 + \\ \left(\overline{v_{n,\,\mathrm{Ref},\,\max}^2} + \overline{v_{n,\,CMP}^2} \right) \end{array} \right\} \le \left(\frac{V_{Ref}}{2^{n+1}} \right)^2, \tag{3.122}$$

considering all mismatches and noise are not correlated.

3.5.2 Design of a High-Speed Comparator

As illustrated in Fig. 3.34, the high-speed comparator for folding and interpolating A/D converter can be constructed by using an *input stage* and a *latch stage*. The input stage is a differential amplifier with active load [71], [93] and the output signals of this input stage A and B drive the connected latch stage via a current mirror. The outputs of the latch stage are already digitized signals, but are usually buffered using another stage in order to get enough current to drive the following digital circuits.

The core components of the latch stage are, except the mirror transistors M3~M6, two *back-to-back connected inverters* M7~M10, which are controlled by two switches, M12 and M13. The transistor M12 is a PMOS transistor and the transistor M13 a NMOS transistor. Thus, these two switches work in paraphase. The working principle of the latch stage is shown in Fig. 3.35.

While the system clock is set to be high, the PMOS switch M12 is turned off in order to save power dissipation and the NMOS switch M13 is turned on to evaluate the information applied by the input stage of the comparator M1~M6. At this moment there is a tiny voltage V_{DS} left between the two terminals V_{OutP} and V_{OutN}. The amount of this voltage V_{DS} can be controlled by input voltage and transistor sizing and the direction of V_{DS} is controlled by the input signals V_{InP} and V_{InN} since the drain and source terminals of the MOS transistor are inherently defined by their potential difference.

In case that the input signal V_{InP} is higher than another input signal V_{InN}, the potential A is lower than the potential B. Hence, the output signal V_{OutP} is higher than the output signal V_{OutN} and the voltage V_{DS} drops from left to right, as drawn as solid line in Fig. 3.35. On the contrary, if the input signal V_{InP} is lower than another input signal V_{InN}, the voltage V_{DS} drops from the right side to the left side, which is drawn as dotted line in Fig. 3.35.

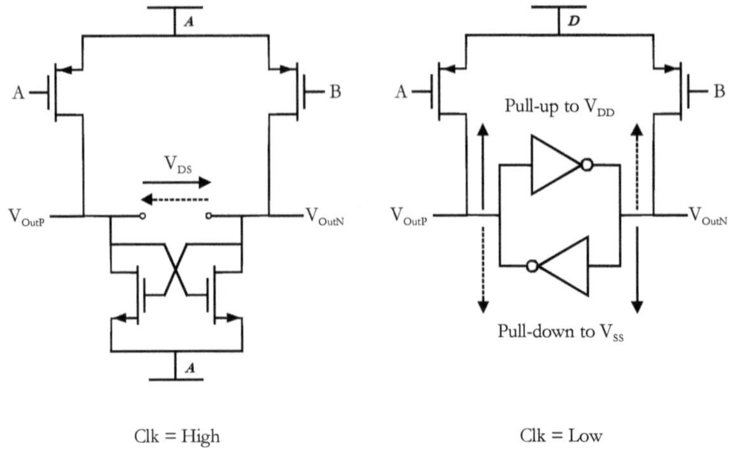

Fig. 3.35: Operating phases of the latch stage

While the clock signal is set to be low, as shown on the right side of Fig. 3.35, the PMOS switch M12 is turned on and the NMOS switch M13 is turned off. The core of the latch stage is changing into two back-to-back connected inverters, which are composed of M7, M9 and M8, M10. The voltage V_{DS}, generated by the input signals V_{InP} and V_{InN} while the clock signal was at high level, drives now the back-to-back connected inverters to V_{DD} and V_{SS}, respectively.

Since the two inverters are back-to-back connected, the switching process carried out by the two inverters is very fast. The output signals, V_{OutP} and V_{OutN}, are pulled up to the positive power supply V_{DD} on the one side and pulled down to the negative power supply V_{SS} on the other side. The output signals are now digitized and are ready to drive the following digital circuits.

The greatest advantage of this kind of comparator is its simplicity and efficiency. The back-to-back connected inverters, suggested by Mr. W. C. Slemmer for bipolar transistors in 1970 [94] and by Mr. H.-L. Fiedler for NMOS transistors in 1981 [95], are more efficient due to the pull-up effect than the back-to-back connected NMOS transistors, suggested in Ref. [93]. Apart from this, the additional PMOS switch M12 helps to save power dissipation.

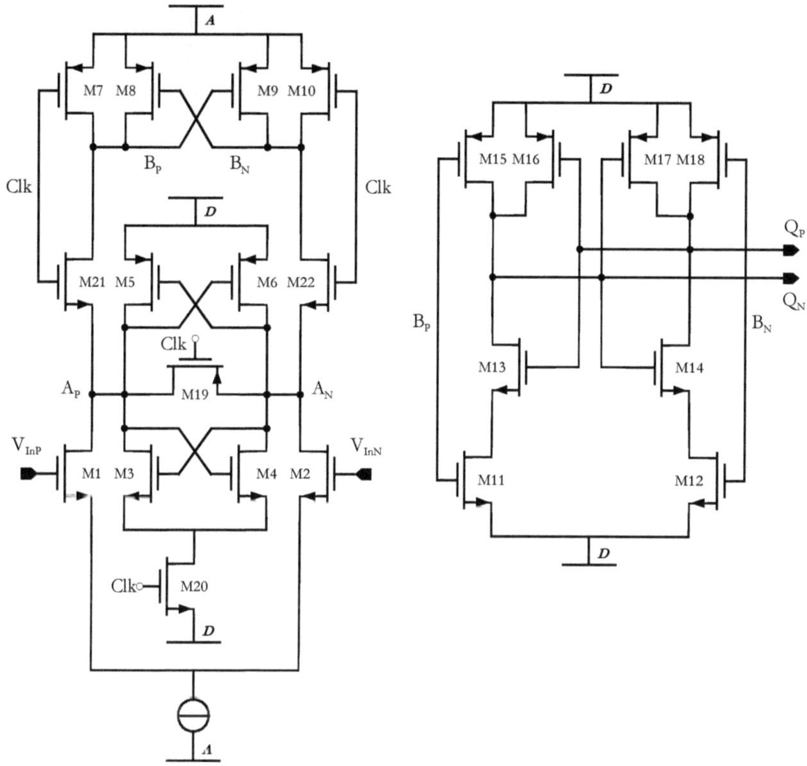

Fig. 3.36: Schematic of low-power comparator

The current mirror between the input stage and the latch stage can be eliminated in order to save power and reduce the die area. If the PMOS active load transistors are merged together with the PMOS transistors in the inverters, then the PMOS switch, as shown in the schematic of Fig. 3.36 (on the left), can be avoided. An extra SR-latch can be added to latch the output signals while the clock is at its low level and then fed into the following digital circuitry, as shown in Fig. 3.36 on the right hand side.

3.5.3 Capacitive Digital Disturbance Inside the Comparator

Although the comparator depicted in Fig. 3.36 performs well with respect to resolution and speed, it can suffer from *crosstalk* caused by digital signals, especially the clock. This can cause unstable behavior, especially near transition regions when used in a flash A/D converter.

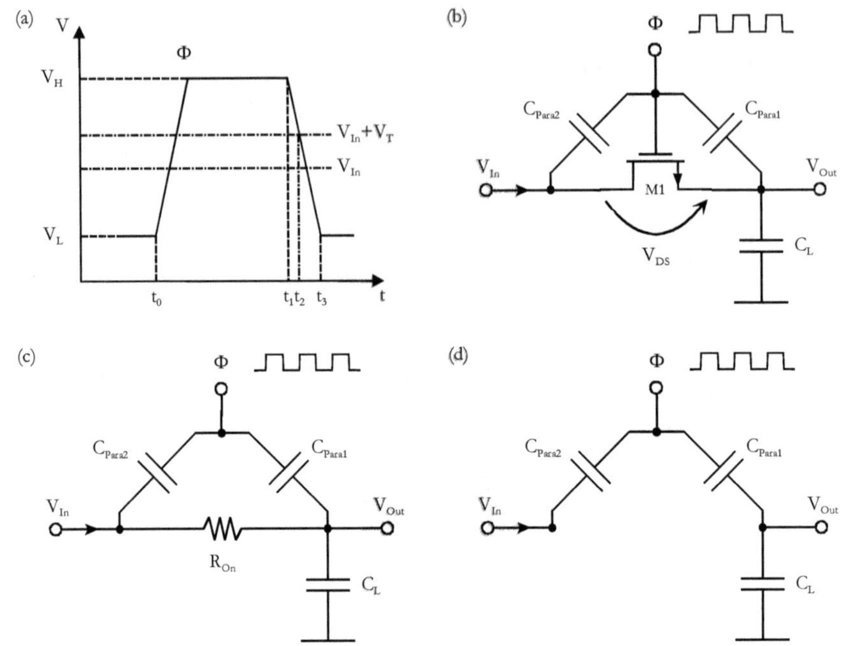

Fig. 3.37: Clock timing diagram and model of the clock feedthrough

Apart from the crosstalk through the substrate and capacitive crosstalk between interconnecting wires, there is also crosstalk caused by the parasitic capacitances and inductors of the wires and components and must be also taken into consideration. The reason that this phenomenon is difficult to be evaluated since the simulations usually employ ideal voltage or current sources as input signals and power supply voltages and require perfect modeling of MOS transistors. Since real sources possess finite source impedance, it is impossible to realize circuit nodes exhibiting either zero or infinite impedances. Such circuit nodes are then susceptible to crosstalk. Especially crosstalk caused by digital circuits presents a formidable problem due to high voltage swing typical for digital circuits. On the other hand, parasitic capacitances of MOS switches are not precisely known and owe nonlinear. The disturbance caused by digital signals can thus cause considerable distortion of analog signals due to the crosstalk while difficult to predict. A good method to simulate these effects is to use real sources, e.g. by inserting a unity-gain amplifier between the ideal source and the simulated circuit, but the problems due to modeling parameters are difficult to combat.

In this section, we will concentrate on the disturbance caused by the switches inside the comparators which represents a major problem here. The problem is named *kickback effect* [93] in the literature. Such kind of crosstalk is also known as *clock feedthrough* (CFT) or *charge injection* [78], and it represents a main challenge for analog discrete-time circuits, e.g. *switched-capacitor* (SC) or *switched-current* (SI) techniques. Some analyses of the charge injection in analog MOS switches can be found in [96, 97]. Methods to compensate the charge injection with *dummy transistors* were proposed by Eichenberger and Guggenbuhl [98].

Consider the configuration of an NMOS switch illustrated in Fig. 3.37 [99], [100]. Fig. 3.37(a) shows the timing diagram of the clock signal, whereas Fig. 3.37(b) shows a model of the clock feedthrough and Fig. 3.37(c) and Fig. 3.37(d) are models while the transistor M1 is turned on or turned off, respectively.

If the transistor M1 is turned on, the transistor M1 acts like a resistor with value of R_{On}. For sufficiently large time constant, V_{Out} is equal to V_{In} and CFT is not a problem. On the contrary, if the transistor M1 is now turned off as shown in Fig. 3.37(d), the effective CFT voltage V_{CFT} can be described as:

$$V_{CFT} \propto -\frac{C_{Paral}}{C_{Paral}+C_L}\cdot\left|\Delta V_{Clk}\right| = -\frac{C_{Paral}}{C_{Paral}+C_L}\cdot\left|V_H - V_L\right| , \tag{3.123}$$

where V_H is the voltage when the clock signal is high and V_L is the voltage when the clock signal is low.

For those cases where the load capacitor C_L is much larger than the parasitic capacitance C_{Paral}, the clock feedthrough effect is limited. But, if the load capacitance is merely a small gate capacitance, compared to the parasitic capacitance, then the disturbance caused by the clock signal is more significant.

However, the model shown in Fig. 3.37(d) does not show a dependence on the input voltage as expected according to Fig. 3.37(b). In reality, the term $(V_H - V_L)$ is reduced, since the charge injection caused by the clock signal during the time t_1 and t_2 is not effective due to the switch M1 still conducting while connected to the input signal V_{In}. Thus, the Eq. (3.123) can be rewritten as:

$$V_{CFT}(V_{In}) = \frac{-C_{Paral}}{(C_L + C_{Paral})}\cdot(V_{In} + V_T - V_L) . \tag{3.124}$$

After the time t_2, the transistor M1 is shut off and the injected charge of the capacitors cannot be conducted away.

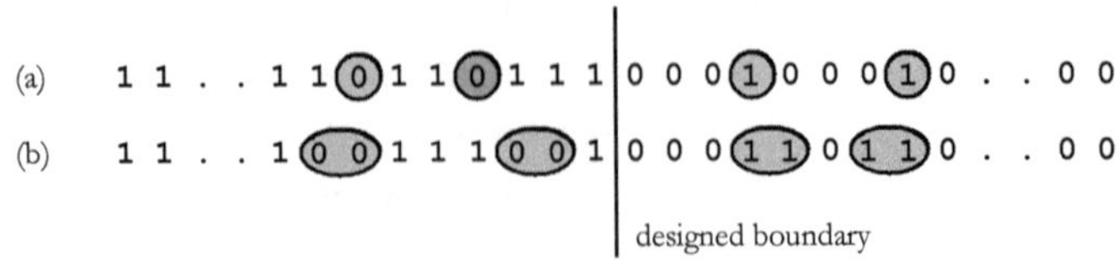

Fig. 3.38: Potential single errors and double errors of the thermometer code

The Eq. (3.124) shows that the disturbance is a function of the input signal V_{In}. Moreover, the CFT is a problem for the clock transition from high to low, but not a problem for the clock transition from low to high. This suggests that any *charge injection compensation* or *charge injection cancellation* is difficult. There are some design techniques that try to reduce the clock feedthrough, but results are disappointing. That is why in this work, we concentrate more on isolation of the clock signal from the input signal rather than on compensating the charge of the capacitors. A new design that accomplishes this will be described in Chapter 4.

3.6 Digital Signal Processing: Error Correction

As discussed above, comparators of the fine converter of the folding and interpolation A/D converter can be designed in a simpler architecture than the comparators designed for a flash A/D converter. The comparators for the flash A/D converter are *"level comparators"*, which compares the input signal to certain voltage levels. Such comparators must have low noise and offset so that special circuit techniques must be employed to achieve this (e.g. *auto-zero techniques*). On the other hand, the input signals of the comparators for the folding and interpolating A/D converter are not comparing the input signal with absolute reference voltages, but merely detecting *"zero-crossings"*, as already discussed in Chapter 3.3.2.

Since the coarse converter of the folding and interpolating A/D converter is kind of flash A/D converter, comparators of the coarse converter of the folding and interpolating A/D converter are level comparators. Thus we would expect that noise and offset of the comparator would be an issue for the coarse converter. But, as the name reveals, the coarse converter is not a converter with high-resolution. For an *m*/*n*-bit folding and interpolating A/D converter, the coarse converter has to resolve only *m* bits, not complete *m*+*n* bits. What critical is at the transition between the neighboring intervals, but

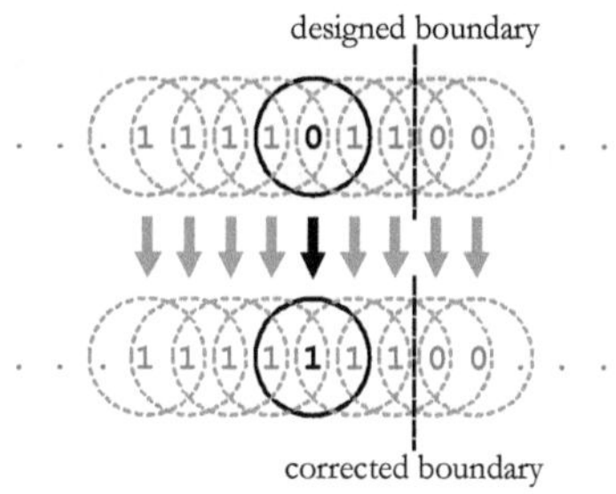

Fig. 3.39: Parallel error correction inside the thermometer code

using proper arrangement in the digital synchronization between the coarse and fine converters can solve this problem. Thus, the noise and offset requirements are greatly reduced.

The flash A/D converter yields the output signal encoded in thermometer code (see Chapter 2.5.1). The ideal thermometer code consists of a series of logic "1"s and a series of logic "0"s of corresponding digital signals. The length of thermometer code is equal to the total number of the comparators. Of course, the length of thermometer code is also equal to the sum of both series of logic "1" and logic "0". For an n-bit converter in which p is the length of the series of logic "1" and q is the length of the series of logic "0", their relationship can be written as:

$$\begin{aligned} &\text{length of thermometer code } = 2^n = p + q\,, \\ &p,\ q \in \mathbb{N} \text{ and } 0 \le p,\ q \le 2^n\ . \end{aligned} \tag{3.125}$$

Supposed that some of the comparators make wrong decision, the series of logic "1" contains some logic 0s, or vice versa, the series of logic "0" will be mixed with logic 1s. In both cases the thermometer code is faulty. The possible *single errors* and *double errors* are illustrated in Fig. 3.38.

The errors found in the thermometer codes represent the degree of susceptibility to noise and interference in the preprocessing stages. The possible error sources include the disturbed reference voltages, the ambiguity of transition regions [101], and mismatch of the comparators, noise etc. Most of these errors in the thermometer codes can be detected and corrected by the digital circuits of the postprocessing stage. Some of the possible correction schemes will be discussed in the following.

A	B	C	Q
0	0	0	0
0	0	1	0
0	1	0	0
1	0	0	0
0	1	1	1
1	0	1	1
1	1	0	1
1	1	1	1

$$Q = AB + BC + CA$$

(a)

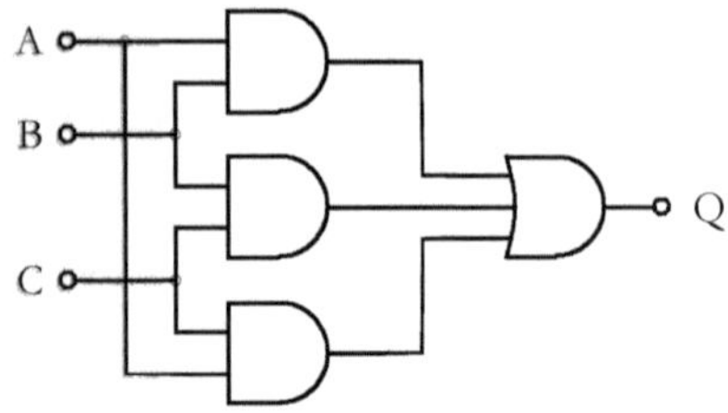

(b)

Fig. 3.40: Truth table of a 3-input voter circuit and corresponding logic circuit

3.6.1 Algorithm for Digital Error Correction

For the sake of simplicity, only errors in the logic "1"-series will be discussed; errors in the logic "0"-series can be treated using the same method. As shown in Fig. 3.39, the single error of the thermometer can be corrected by comparing each comparator output with its neighbors. That is, each comparator output is compared with its two nearest neighbors and the result will replace the original signal. Assuming that there is an error code "0", printed in black in Fig. 3.39, among the "1"-series, the signal will be corrected to "1" after the processing. Such kind of comparison process can be very efficient because all signals are compared in parallel.

This digital error correction can be implemented by means of a *voter circuit* [102]. The voter circuit has odd number of digital input signals and its output signal is decided by the majority, hence the name, of the input signals. The truth table of a 3-input voter circuit and its corresponded logic circuit

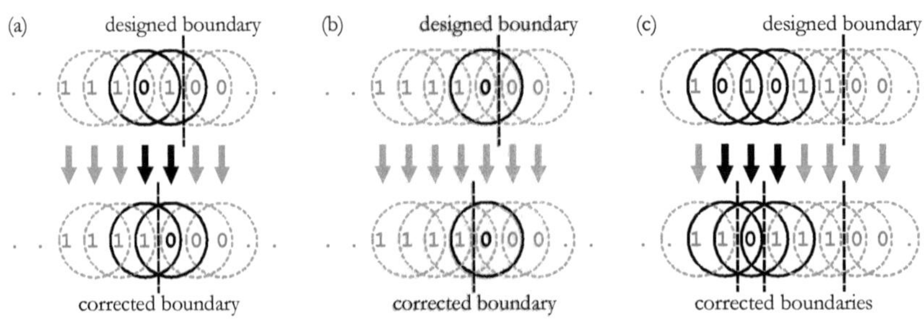

Fig. 3.41: Uncorrectable single errors by a 3-input voter circuit

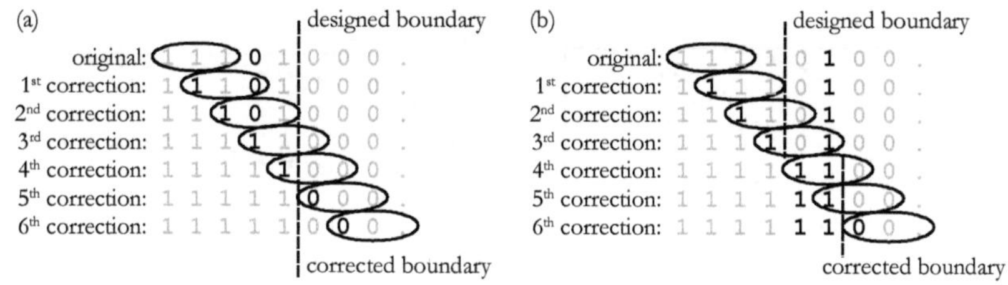

Fig. 3.42: Serial error correction in the thermometer code

are depicted in Fig. 3.40.

Supposing that a single error is located very close to the designed boundary between the logic 1's and 0's series, the corrected boundary will not be equivalent to the designed boundary, as illustrated in Fig. 3.41(a) and Fig. 3.41(b).

This leads to an uncorrectable error of 1 *LSB* if this erroneous thermometer code belongs to the fine converter. Hence, the voter circuit with only 3 inputs cannot correct errors, which are located two digits away from the designed boundary. This is not very favorable for the folding and interpolating A/D converter because the possibility that errors occur in the digits near the designed boundary between logic "0" and "1" series are higher than possibility that errors occur in the digits from other region.

Another case of uncorrectable single errors by a 3-input voter circuit are single errors, which are located next to another, as depicted in Fig. 3.41(c). This kind of single errors leads to two additional boundaries between logic "0" and "1" series in the output signals. But the new generated boundaries can be eliminated by a second stage of 3-input voter circuit. Furthermore, the 3-input voter circuit cannot correct all kinds of double errors and triple errors etc.

Besides the algorithm of *parallel error correction* as shown above, the algorithm of *serial error correction* shown in Fig. 3.42(a) can be another solution. As it can be seen in the example, the source thermometer code has a "0" error at the 4th digit. The digital error correction is carried out from the left side of thermometer code towards the right side. Encircled signals are processed by the 3-input voter circuit, which gives its output for the next correction (printed in black). After the 3rd serial correction the "0" error is corrected. The corrected boundary is matched with the designed boundary, even if the error occurs one digit next to the designed boundary as shown in Fig. 3.42(a). Using serial processing, double single errors like shown in the example in

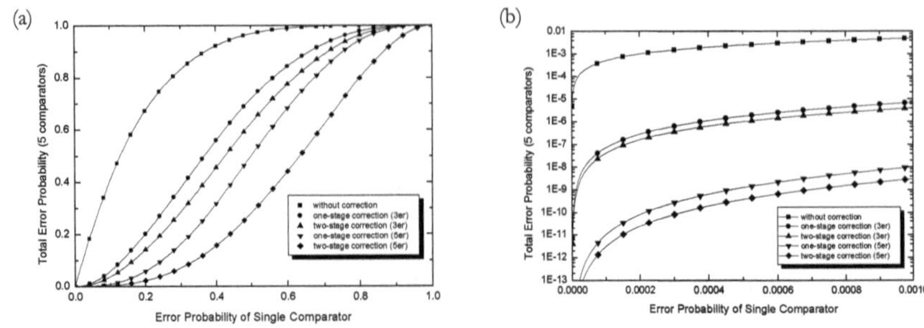

Fig. 3.43: Total error probability of 5 comparators

Fig. 3.41(c) can be eliminated. But, if the single error occurs on the other side of designed boundary, as the example provided in Fig. 3.42(b), the corrected boundary would be displaced by two digits from the designed boundary. Moreover, the serial error correction causes enormous latency compared to the parallel error correction. A 64-digit thermometer code needs a total of 62 processing steps of serial error correction. This means that the serial error correction is not practical.

3.6.2 Multistage Digital Error Correction

Since the 3-input voter circuit cannot solve problems such as double or triple errors in the thermometer code, further investigations are necessary. These must, however, also take hardware amount and latency into consideration. A *two-stage 5-input voter circuit* has been investigated in this book. The principle of a 5-input voter circuit is similar to that of a 3-input voter circuit, only the truth table and the implemented logic circuit are more complicated.

Assuming the situations that errors occur in the "1"-series of the thermometer code and all other signals are faultless outside the considering region, the possible states and their corresponding corrections by the 3-input and 5-input voter circuits are summarized in Table 3.3. The cells marked with "✓" are already faultless.

The last state, numbered 31, was originally faultless. Therefore, it should not be taken into account. Supposing the probability that a comparator makes an error is equal to p, the total error probabilities of the 5 comparators is

$$f_{Original}(p) = p^5 + 5p^4 \cdot (1-p) + 10p^3 \cdot (1-p)^2 + 10p^2 \cdot (1-p)^3 + 5p \cdot (1-p)^4 , \tag{3.126}$$

and

$$p \in R, \ \ 0 \le p \le 1 . \tag{3.127}$$

The total error probability that the result after the error correction has not been properly corrected can be described by summing probabilities of the unchecked cells in Table 3.3. This leads to,

$$f(p)_{3er\_1st} = p^5 + 5p^4 \cdot (1-p) + 10p^3 \cdot (1-p)^2 + 7p^2 \cdot (1-p)^3 , \tag{3.128}$$

$$f(p)_{3er\_2nd} = p^5 + 5p^4 \cdot (1-p) + 10p^3 \cdot (1-p)^2 + 4p^2 \cdot (1-p)^3 , \tag{3.129}$$

$$f(p)_{5er\_1st} = p^5 + 5p^4 \cdot (1-p) + 10p^3 \cdot (1-p)^2 , \tag{3.130}$$

and

$$f(p)_{5er\_2nd} = p^5 + 5p^4 \cdot (1-p) + 3p^3 \cdot (1-p)^2 . \tag{3.131}$$

Depicted in Fig. 3.43(a) is the plot of five estimates for the above formula with p ranging from 0 to 1. As can be seen from the plot, the error correction algorithm has a very good performance while p is small. Varying the error probability of single comparator from 0 to 0.001 as plotted in Fig. 3.43(b) shows that a two-stage error correction with 5 signal inputs can reduce the total error probability to 10^{-6} when compared to that of uncorrected case for $p = 0.001$.

3.7 Digital Signal Processing: Algorithm for Binary Coding

An efficient way to transform the thermometer code into the binary code is to transform the thermometer code into the *1-of-N code* at first. The 1-of-N code marks the boundary between the logic "0"-series and the logic "1"-series with a "1" and sets all other digits of the thermometer code a logic "0". This kind of coding can be implemented easily by using an array of XOR-gates as shown in Fig. 3.44.

Table 3.3: Possible states and results of the corresponding two-stage correction

No.	5 Digit inputs	Probability	3er voter circuit After 1st correction	3er voter circuit After 2nd correction	5er voter circuit After 1st correction	5er voter circuit After 2nd correction
0	00000	p^5	00000	00000	00000	00000
1	00001	$p^4 \cdot (1-p)$	00001	00001	00001	00001
2	00010	$p^4 \cdot (1-p)$	00001	00001	00001	00001
3	00011	$p^3 \cdot (1-p)^2$	00011	00011	00011	00011
4	00100	$p^4 \cdot (1-p)$	00000	00000	10001	10001
5	00101	$p^3 \cdot (1-p)^2$	00011	00011	10011	✓
6	00110	$p^3 \cdot (1-p)^2$	00111	00111	11011	✓
7	00111	$p^2 \cdot (1-p)^3$	00111	00111	✓	✓
8	01000	$p^4 \cdot (1-p)$	10000	10000	10000	10000
9	01001	$p^3 \cdot (1-p)^2$	10001	10001	10011	✓
10	01010	$p^3 \cdot (1-p)^2$	10101	11011	11011	✓
11	01011	$p^2 \cdot (1-p)^3$	10111	✓	✓	✓
12	01100	$p^3 \cdot (1-p)^2$	11100	11100	11011	✓
13	01101	$p^2 \cdot (1-p)^3$	✓	✓	✓	✓
14	01110	$p^2 \cdot (1-p)^3$	✓	✓	✓	✓
15	01111	$p \cdot (1-p)^4$	✓	✓	✓	✓
16	10000	$p^4 \cdot (1-p)$	10000	10000	10000	10000
17	10001	$p^3 \cdot (1-p)^2$	10001	10001	10001	10001
18	10010	$p^3 \cdot (1-p)^2$	10001	10001	11001	✓
19	10011	$p^2 \cdot (1-p)^3$	10011	10011	✓	✓
20	10100	$p^3 \cdot (1-p)^2$	11000	11000	11001	✓
21	10101	$p^2 \cdot (1-p)^3$	11011	✓	✓	✓
22	10110	$p^2 \cdot (1-p)^3$	✓	✓	✓	✓
23	10111	$p \cdot (1-p)^4$	✓	✓	✓	✓
24	11000	$p^3 \cdot (1-p)^2$	11000	11000	11000	11000
25	11001	$p^2 \cdot (1-p)^3$	11001	11001	✓	✓
26	11010	$p^2 \cdot (1-p)^3$	11101	✓	✓	✓
27	11011	$p \cdot (1-p)^4$	✓	✓	✓	✓
28	11100	$p^2 \cdot (1-p)^3$	11100	11100	✓	✓
29	11101	$p \cdot (1-p)^4$	✓	✓	✓	✓
30	11110	$p \cdot (1-p)^4$	✓	✓	✓	✓
31	11111	$(1-p)^5$	✓	✓	✓	✓

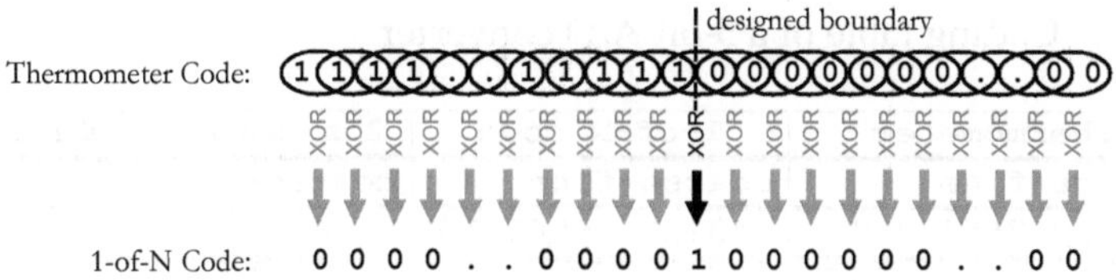

Fig. 3.44: Transformation from thermometer code to 1-of-N code

Certainly, if there were still uncorrected errors in the thermometer code, these errors would be also processed by the 1-of-N code. But, according to the above discussion, the probability is quite low after two-stage error correction.

Moreover, the *LSB* of the coarse converter can be directly determined by the thermometer code of the fine converter. Normally, the *LSB* of the coarse converter is determined by the thermometer code of the coarse converter itself. The *least significant digit* (LSD) in thermometer code of the fine converter alternates its value from "0" to "1" or vice versa while the input signal of the whole A/D converter changes between the neighboring intervals. The *LSB* of the coarse converter has the same function. Its value also alternates between "0" and "1" while the input signal of the A/D converter changes between the neighboring intervals. Therefore, the *LSB* of the coarse converter can be deduced from the last significant digit of the thermometer code of the fine converter directly, without coding the results of the comparators of the coarse converter. Thus, such a mechanism synchronizes the *LSB* of the coarse converter with the fine converter. Considering a simple example of a 5-bit converter, constructed from a 2-bit coarse converter and a 3-bit fine converter, the relationship among the thermometer codes, the 1-of-N codes, and their binary coding can be clearly deduced.

A 5-bit converter exhibits 32 states in total, as can be seen in Table 3.4. Since the resolution of the fine converter in this case is 3-bit, the length of its thermometer code is 8. In addition, the length of the thermometer code from a 2-bit coarse converter is 5, so that the function of over-range indication is also included. The states, marked by "<<" and ">>" in Table 3.4, indicate that the input signal is outside the designed range. In both cases, the contents of fine thermometer codes are not of interest. Therefore, their digits are marked with "X". Nevertheless, during the transformation from the thermometer code to the 1-of-N code, there is one digit eliminated from the thermometer code. The left-most digit (*most significant digit*, MSD) of the 1-of-N code is also marked with "X", as can be seen in the table. The zeros in the ther-

Table 3.4: Coding table of a 5-bit A/D converter

	Thermometer	1-of-N code	2's Comple.	Decimal
	coarse-fine	coarse-fine	coarse-fine	
No.	43210-76543210	43210-76543210	43-210	
<<	00000-XXXXXXXX	00000-XXXXXXXX	XX-XXX	<<
1	00001-00000001	X0001-X0000001	10-000	16
2	00001-00000011	X0001-X0000010	10-001	17
3	00001-00000111	X0001-X0000100	10-010	18
4	00001-00001111	X0001-X0001000	10-011	19
5	00001-00011111	X0001-X0010000	10-100	20
6	00001-00111111	X0001-X0100000	10-101	21
7	00001-01111111	X0001-X1000000	10-110	22
8	00001-11111111	X0001-X0000000	10-111	23
9	00011-11111110	X0010-X0000001	11-000	24
10	00011-11111100	X0010-X0000010	11-001	25
11	00011-11111000	X0010-X0000100	11-010	26
12	00011-11110000	X0010-X0001000	11-011	27
13	00011-11100000	X0010-X0010000	11-100	28
14	00011-11000000	X0010-X0100000	11-101	29
15	00011-10000000	X0010-X1000000	11-110	30
16	00011-00000000	X0010-X0000000	11-111	31
17	00111-00000001	X0100-X0000001	00-000	0
18	00111-00000011	X0100-X0000010	00-001	1
19	00111-00000111	X0100-X0000100	00-010	2
20	00111-00001111	X0100-X0001000	00-011	3
21	00111-00011111	X0100-X0010000	00-100	4
22	00111-00111111	X0100-X0100000	00-101	5
23	00111-01111111	X0100-X1000000	00-110	6
24	00111-11111111	X0100-X0000000	00-111	7
25	01111-11111110	X1000-X0000001	01-000	8
26	01111-11111100	X1000-X0000010	01-001	9
27	01111-11111000	X1000-X0000100	01-010	10
28	01111-11110000	X1000-X0001000	01-011	11
29	01111-11100000	X1000-X0010000	01-100	12
30	01111-11000000	X1000-X0100000	01-101	13
31	01111-10000000	X1000-X1000000	01-110	14
32	01111-00000000	X1000-X0000000	01-111	15
>>	11111-XXXXXXXX	00000-XXXXXXXX	XX-XXX	>>

mometer code and 1-of-N code are printed with gray color to emphasize the periodicity in the code words.

It can be recognized in this table that through proper definition of the beginning value of the 1st state, the *LSB* of the coarse converter in the 2's complement code (bit 3) is inverted from the last digit in the fine thermometer code (bit 0). Both columns are printed in bold type in Table 3.4.

3.8 Subsystem: Synchronization

3.8.1 Introduction

The folding and interpolating A/D converter must be synchronized both at the analog front-end and/or at the digital rear-end as can be seen in Fig. 2.19.

The coarse and the fine converter work in parallel behind the sampling stage (Fig. 2.19). The coarse converter, which can be constructed in a flash architecture without circuits causing various interstage delays, has usually much lower latency than the fine converter, which has a longer latency due to interstage delays. If the latency difference between both converters cannot be compensated for an *m*/*n*-bit folding and interpolating A/D converter, there would be inconsistent errors, value of which can be up to 2^n *LSB*, and which occur at the interval edges (Fig. 2.20). Hence, the synchronization stage cannot be avoided.

If the input sides of both converters are fed from the output of the THA stage, the input sides of both converters are synchronized. This kind of synchronization can be called as *analog synchronization*.

However, the analog synchronization cannot compensate for the signal racing between the two converters completely. To avoid undesired signal racing in the converters, another mechanism for synchronization, for example *digital synchronization*, is necessary. Using the same clock signal from the THA stage for the outputs stages of the converters enables the synchronization at the converter output [48].

Since the THA stage has already been treated above, the discussion in this section will concentrate only on the digital synchronization stage.

3.8.2 Algorithm of Digital Synchronization and Implementation

An example of digital converter synchronization has been proposed by Mr. Flynn [47, 70] and it is shown in Fig. 3.45. Let us consider again the 5-bit A/D converter introduced in Chapter 3.7 and Table 3.4. In this example, the most significant bit *MSB* is the 4th bit in the 2's complement code (Table 3.4). We can denote the 3rd bit in the 2's complement code as MSB_{-1} thus representing the second significant bit and the 2nd bit in the 2's complement code as MSB_{-2}, which stands for the third significant bit of the entire converter. All signals

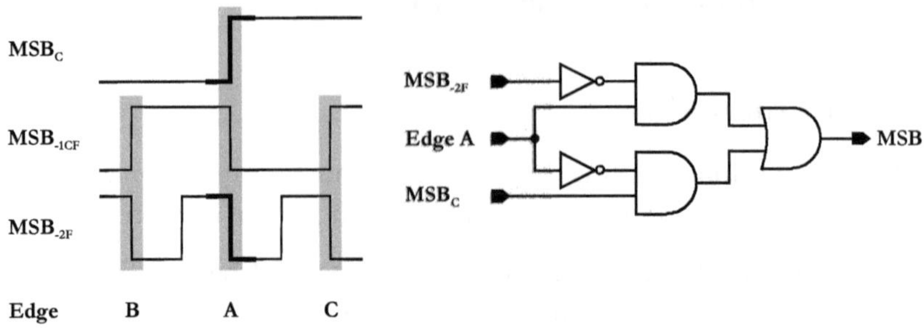

Fig. 3.45: General principle of the digital synchronization

MSB, *MSB*$_{-1}$, and *MSB*$_{-2}$ mentioned here represent the final synchronized signals.

The signals *MSB*$_C$ and *MSB*$_{-1CF}$ in Fig. 3.45 are the output signals of the coarse converter and the signal *MSB*$_{-2F}$ in Fig. 3.45 is the output signal of the fine converter. As discussed in Chapter 3.7, the signal *MSB*$_{-1CF}$ can be deduced from the thermometer code of the fine converter. Thus, the signal *MSB*$_{-1CF}$ is already synchronized with the fine converter. Only the signals *MSB*$_C$ and *MSB*$_{-2F}$ have to be synchronized. Hence the critical time at which the synchronization must take place is given by the edge A, which is drawn in gray color on the left side of Fig. 3.45.

Considering the signals *MSB*$_C$ and *MSB*$_{-2F}$, the logic circuit for its computation is shown on the right side of Fig. 3.45. This logic circuit can be explained as follows:

> If the signal *MSB*$_C$ is at the critical edge A, then the signal *MSB* is decided by the signal *MSB*$_{-2F}$ and is equal to the inversion of the signal *MSB*$_{-2F}$. Otherwise, if the signal *MSB*$_C$ is not at the critical edge A, then the signal *MSB* is equal to its original signal *MSB*$_C$.

Since the signal *MSB*$_{-1CF}$ is deduced from the fine converter, the synchronization between the signal *MSB*$_C$ and fine converter can also be carried out by using the signal *MSB*$_{-1CF}$, instead of using the signal *MSB*$_{-2F}$.

The first aid is to detect if the input signal of the A/D converter is in the critical regions, which are drawn with gray thick line in Fig. 3.45. The critical regions can be detected by the thermometer codes of the coarse and the fine A/D converter.

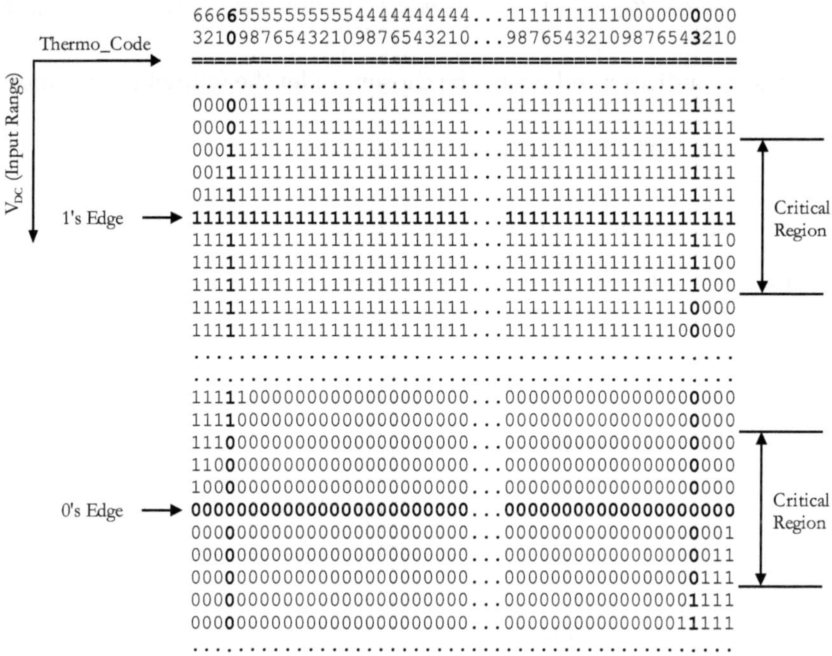

Fig. 3.46: Variable width definition of the critical region

The ith edge locates at the right side of the $(i-1)^{th}$ edge and at the left side of the $(i+1)^{th}$ edge. It can be described as follows:

$$Edge_i = (Coarse\_TCode_{i-1}) \text{ AND } (Coarse\_NTCode_{i+1}) \, , \tag{3.132}$$

and

$$i = 2 \cdot j \qquad 1 \le j \le \left(2^{m-1} - 1\right) \qquad j \in \mathbb{N} \, , \tag{3.133}$$

where m is the resolution of the coarse converter.

The term $Coarse\_TCode_i$ means the ith positive thermometer code of the coarse A/D converter and the term $Coarse\_NTCode_i$ indicates the ith negative thermometer code of the coarse A/D converter.

The width of the edge line (critical region) can be defined by the thermometer codes of the fine converter. Fig. 3.46 is an example, which shows how the critical region of a 6-bit fine A/D converter is defined.

All critical regions of the input signal range can be described as follows:

$$Critical\_Region = (Fine\_TCode_x) \text{ XNOR } (Fine\_TCode_y)\,, \tag{3.134}$$

whereas the indices x and y must be chosen under the following conditions,

$$x + y = 2^n - 1 \qquad x,\, y \geq 1 \qquad x,\, y \in \mathbb{N}\,, \tag{3.135}$$

where n is the resolution of the fine converter.

Same as before, the $Fine\_TCode_x$ indicates the xth positive thermometer code of the fine A/D converter. In Fig. 3.46, x is to be set to 3, without loosing generality, and y is equal to 60.

Besides, if there are only even-numbered edges to be considered, the equation can be written as:

$$\begin{aligned} Critical\_Region &= (Fine\_TCode_x) \text{ NOR } (Fine\_TCode_y) \\ &= \overline{(Fine\_TCode_x)} \text{ AND } \overline{(Fine\_TCode_y)}\,, \end{aligned} \tag{3.136}$$

the constraints for the indices x and y are the same with those mentioned above.

Fig. 3.47 illustrates the digital signal flowing chart and the system clock distribution of the error correction, the binary coding, and the synchronization stage. The realized logic for edge detection and synchronization can be found in the implementation part of this book.

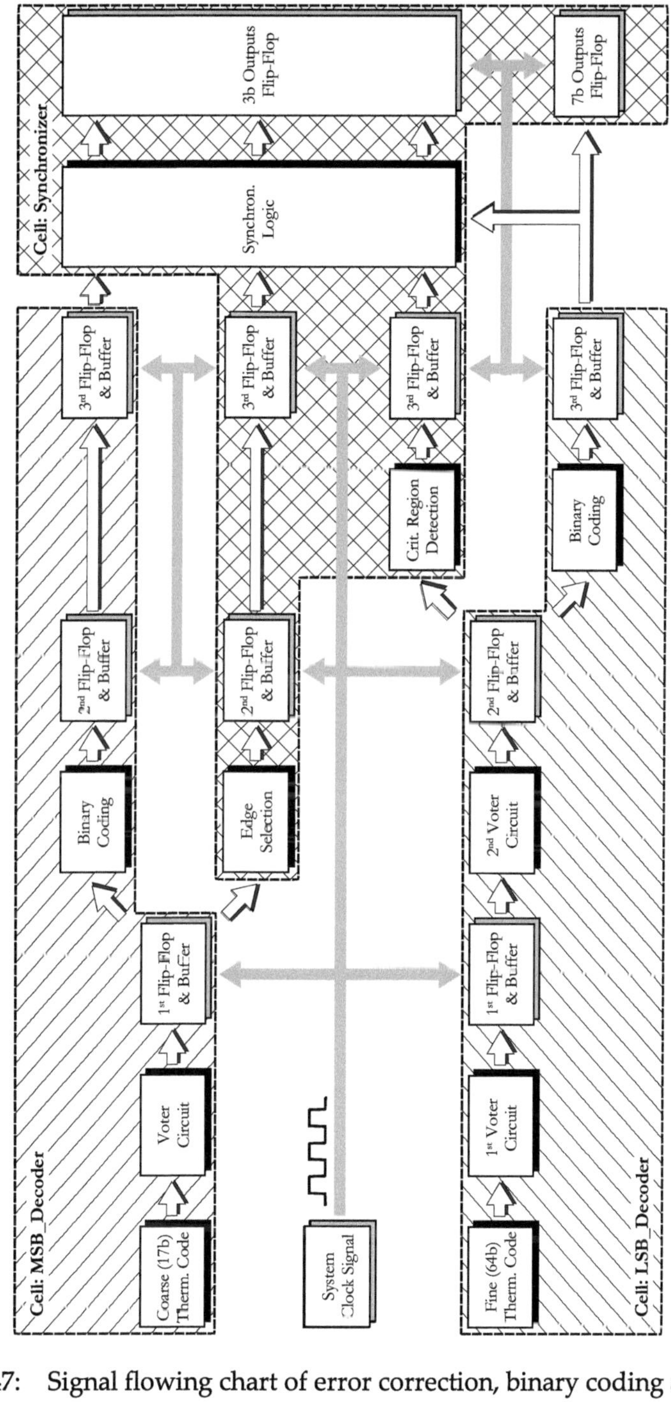

Fig. 3.47: Signal flowing chart of error correction, binary coding and synchronization

CHAPTER 4

ARCHITECTURE AND DESIGN OF CMOS FOLDING AND INTERPOLATING A/D CONVERTERS

Epitome

4.1 Introduction

In this chapter, architectures for low-power, high-speed CMOS folding and interpolating A/D converters are investigated. Since various combinations of different coarse and fine resolutions yield high modularity of this type of converter, different folding and interpolating resolutions are possible and lead to different results. As an example a 10-bit folding and interpolating A/D converter is investigated for later implementation. Based on application, the power dissipation and the die area are of main interest. Therefore, estimates of the power dissipation and the die area play a prominent role in the investigation.

There are no commercially available models for A/D or D/A converters, although attempts have been made to build macromodels using e.g. *SPICE* (*Simulation Program with Integrated Circuit Emphasis*) by numerous researchers. This is due to the difficulty in modeling their AC performance [103]. Besides, for the high-speed applications, the package substrate and parasitics should also be included, but their parameters are difficult to obtain.

4.1.1 Micromodel and Macromodel

What an IC designer obtains from chip manufacturers are models and their parameters of transistors and other passive components, such as of capacitors and resistors. Simulations based on such models can be designated as *micromodel* simulation [103].

Some computer aided design (CAD) tools allow the extraction of the parasitic passive components from the layout and an inclusion in the *post-layout simulation*. Since the total number of parasitic components may be huge and most of them are of tiny values in practice, the nondominant parasitic components should be sorted out at first. Finally, the remaining parasitic passive components can be added to the schematic and re-simulated in order to account for effects of the parasitic passive components [104].

This kind of simulation is still based on the micromodel simulation, but it is more realistic. Therefore, it can be named as *improved micromodel* simulation. The simulation itself is more demanding than pure micromodel simulation.

Contrary to the micromodel simulation, the *macromodel* simulation treats a complete A/D converter as a single component, like a "black box". Thus only a limited number of parameters of the converter at the system level are considered. The macromodel simulation is particularly suitable at system level, e.g. to determine and optimize an architecture of an A/D converter. The macromodel features fast simulation times and can be easily modified. But, on the other side, the macromodel simulation is not precise enough to characterize sufficiently the component performance in all detail and include nonideal effects, such as parasitics. Thus, trade-off between the micromodel and the macromodel must be found to improve the prediction of performance of the simulated component before the costly simulation at the transistor level can commence.

4.1.2 Boyle Model and ADSpice Model

Besides the micromodel simulation and the macromodel simulation discussed above, modeling of analog building blocks, for example operational amplifiers, can be also useful. As an example, the *Boyle model*, introduced by Mr. Boyle in 1974 [105], describes the behavior of operational amplifier. However, this model considers only two frequency-shaping poles and no zeros, and cannot describe performance of operational amplifiers with sufficient accuracy [103].

Contrary to the Boyle model, the *ADSpice model* has an open architecture, which allows unlimited numbers of poles and zeros and can simulate this type of circuits more accurately, especially in respect to AC and transient simulation [103]. The ADSpice model contains three main segments: the input and gain stage, the pole/zero stages, and the output stage. This concept of segmenting the target object into several controllable blocks will be used to model the folding and interpolating A/D converter in this work.

Based on the highly modular characteristic of the folding and interpolating A/D converter, a new middle class model will be developed that allows to estimate the performance, such as die area or power dissipation, of the conversion system. Finally, the pros and cons of the different architectures of folding and interpolating A/D converters will be discussed.

4.2 Architecture Optimization

4.2.1 Introduction

In this chapter, architecture of high-speed folding and interpolating A/D converters, which are suitable for low-power embedded systems, is investigated. The generic architecture will be based on the block diagram shown in Fig. 2.19 and its components described in Chapter 3. A 10-bit A/D converter operating at 20 MHz clock frequency will be used as an experimental vehicle.

Above all, the resolutions of the coarse and the fine converters, including the folding resolution and the interpolating resolution, will be looked into in detail because they decidedly affect the power dissipation and the die area. The optimization will be based on a primary hard scaling model as discussed below [48].

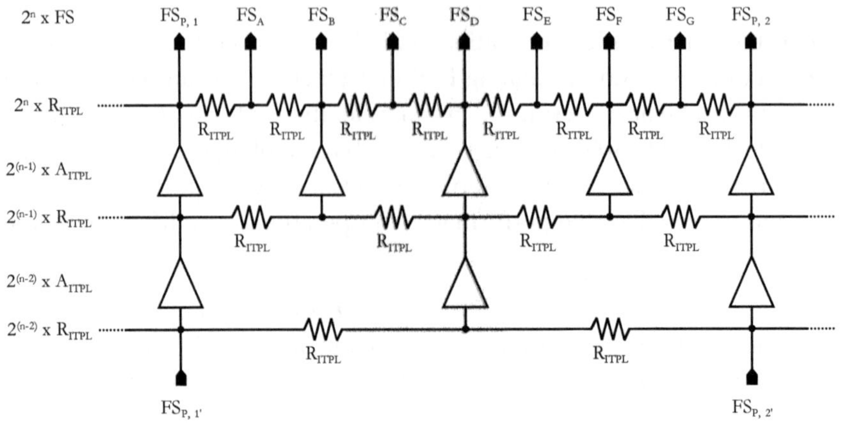

Fig. 4.1: Estimate of components of the cascaded 3-stage, 2-times interpolation

4.2.2 Die Area Analysis

Although the total number of comparators used in the folding and interpolating A/D converter is much lower than in the flash A/D converter, the folding and interpolating A/D converter still requires a considerable amount of hardware and has an impact on the total power dissipation and the die area. Thus, investigations must be made in order to be able to optimize the performance of the conversion system. For the sake of discussion, the analysis is confined to the considerations of the die area. Similar analysis, however, can be used for the power dissipation.

4.2.2.1 Comparator Stages

Assuming that the coarse and the fine converter have m-bit and n-bit resolution, respectively, and the comparators designed for both converters can be considered as the identical under the first order of approximation, the total die area of the comparator stage can be described as follows:

$$\left[\left(2^m+1\right)+2^n\right]\cdot S_{CMP}\ , \tag{4.1}$$

whereas the term S_{CMP} indicates the die area of a single comparator. The coarse converter has function of out-of-range indicator, hence, the total number of comparators is equal to the total representable states plus 1, instead of minus 1.

4.2.2.2 Interpolation Stage

The next consideration is the interpolation stage. The main components of the cascaded interpolation are the interpolation resistors and the inter-stage amplifiers as discussed in Chapter 3.3.5. A schematic of a cascaded 3-stage, 2-times interpolation stage is based on principle shown in Fig. 3.24 is shown in more detail in Fig. 4.1.

As depicted in Fig. 4.1, a fine converter of n-bit resolution needs 2^n differential folding signals to be fed into its 2^n comparators. Thus, the last stage of the interpolation requires 2^n interpolating resistors R_{ITPL} and $2^{(n-1)}$ inter-stage amplifiers to meet its need. The number of components for the preceding stages can be obtained by the same method. Hence, the total die area of the interpolating resistors used for an n-bit fine converter with n_1-bit folding resolution and n_2-bit interpolating resolution in cascaded multi-stage, 2-times interpolating structure can be described as:

$$\left[2\cdot\left(2^{n_1}\cdot 2+2^{n_1}\cdot 2^2+\ldots+2^{n_1}\cdot 2^{n_2}\right)\right]\cdot S_{R\_ITPL}$$
$$=\left[2\cdot\left(2^{n_1}\cdot\sum_{i=1}^{n_2}2^i\right)\right]\cdot S_{R\_ITPL}=\left[2^{(n_1+1)}\cdot\sum_{i=1}^{n_2}2^i\right]\cdot S_{R\_ITPL}, \tag{4.2}$$

where the first factor 2 is because of the differential signals and the term S_{R_ITPL} indicates the real die area of the interpolating resistor.

Moreover, the number of amplifiers can be described as:

$$\left(2^{n_1}\cdot 2+2^{n_1}\cdot 2^2+\cdots+2^{n_1}\cdot 2^{(n_2-1)}\right)\cdot S_{A\_ITPL}$$
$$=\left[2^{n_1}\cdot\sum_{i=1}^{n_2-1}2^i\right]\cdot S_{A\_ITPL}, \tag{4.3}$$

the term S_{A_ITPL} shows the die area of an inter-stage amplification.

Since folding signals are, as illustrated in Fig. 3.20, differential signals with 180° phase difference, signals can be interpolated between the last positive folding signal and the first negative folding signal or between the last negative folding signal and the first positive folding signal. As a result, all folding signals, both the input signals and the output signals of the interpolation stage, are chained onto a circle like shown in Fig. 3.13 and Fig. 3.20. The shifted folding signals and the interpolated folding signals are distributed equidistantly and all amplifier outputs are loaded symmetrically and equally.

4.2.2.3 Folding Stage

As shown in Fig. 3.11, the basic module in the folding stage is the folding amplifier, which contains cascaded amplifiers as it can be seen in Fig. 3.14. Given the folding amplifier as the basic unit for a folding and interpolating A/D converter with m-bit coarse resolution, the mandatory number of folding amplifiers is 2^m. In order to avoid the signal distortion at the both ends of the input signal range and to add an offset DC signal to the output signal of the folding stage, three additional folding amplifiers are to be attached to the folder. Furthermore, for a fine converter with n_1-bit folding resolution, the total number of folding amplifiers can be summarized as:

$$2^{n_1} \cdot (2^m + 3) \cdot S_{FA} \text{ ,} \tag{4.4}$$

where the term S_{FA} indicates the die area of a folding amplifier. Besides, based on the experience made during the chip layout, it can be assumed that the die area of the resistor loads inside a folder is approximately the same as that of the folding amplifier. Therefore, the Eq. (4.4) can be rewritten as:

$$2^{n_1} \cdot (2^m + 4) \cdot S_{FA} \text{ ,} \tag{4.5}$$

for the whole folding stage.

The number of reference resistors is equal to the number of folding amplifiers without considering the generation of the offset DC signal. Hence, the die area of the reference resistors can be expressed as:

$$2^{n_1} \cdot (2^m + 2) \cdot S_{Ref} \text{ ,} \tag{4.6}$$

where the term S_{Ref} represents the layout die area of a single resistor.

4.2.2.4 Complete Conversion System

Consequently, the core die area of a folding and interpolating A/D converter in *m*/*n*-architecture can be summarized as:

$$\left[2^{n_1}\cdot(2^m+2)\cdot S_{Ref}\right]+\left[2^{n_1}\cdot(2^m+4)\cdot S_{FA}\right]+\left[2^{(n_1+1)}\cdot\sum_{i=1}^{n_2}2^i\right]\cdot S_{R\_ITPL}+$$
$$\left[2^{n_1}\cdot\sum_{i=1}^{n_2-1}2^i\right]\cdot S_{A\_ITPL}+\left[(2^m+1)+2^n\right]\cdot S_{CMP}\;. \tag{4.7}$$

If the coarse comparators and the fine comparators are treated as different components, then Eq. (4.7) can be given more specifically as:

$$\left[2^{n_1}\cdot(2^m+2)\cdot S_{Ref}\right]+\left[2^{n_1}\cdot(2^m+4)\cdot S_{FA}\right]+\left[2^{(n_1+1)}\cdot\sum_{i=1}^{n_2}2^i\right]\cdot S_{R\_ITPL}+$$
$$\left[2^{n_1}\cdot\sum_{i=1}^{n_2-1}2^i\right]\cdot S_{A\_ITPL}+(2^m+1)\cdot S_{CMP\_C}+2^n\cdot S_{CMP\_F}\;. \tag{4.8}$$

The interpolation stage has two variables, S_{R_ITPL} and S_{A_ITPL}, which cause the estimate of the die area to be rather complicated. An alternative is to estimate the ratio between S_{R_ITPL} and S_{A_ITPL} from the layout considerations and to eliminate one of the two variables to simplify the modeling.

Considering the technology used for implementation of 10-bit A/D converter (Chapter 4.3.2), a single CMOS amplifier has a width of $41.4\ \mu\text{m}$ and a height of $47.2\ \mu\text{m}$, which covers an area of $1954.08\ \mu\text{m}^2$. All three resistor ladders, composed of 224 resistors, have a width of $670\ \mu\text{m}$ and heights of $21.7\ \mu\text{m}$, $24\ \mu\text{m}$, and $70\ \mu\text{m}$, respectively. The single resistor is exhibits an area of $346\ \mu\text{m}^2$, which leads to the following relation:

$$\begin{aligned} S_{R\_ITPL} &= 0.177\cdot S_{A\_ITPL} \\ &\cong 0.2\cdot S_{A\_ITPL}\;. \end{aligned} \tag{4.9}$$

Eq. (4.8) can be rewritten as:

$$2^{n_1} \cdot \begin{bmatrix} (2^m + 2) \cdot S_{Ref} + (2^m + 4) \cdot S_{FA} + \\ (1.8 \cdot 2^{n_2} - 2.8) \cdot S_{A\_ITPL} \end{bmatrix} + \tag{4.10}$$
$$(2^m + 1) \cdot S_{CMP\_C} + 2^n \cdot S_{CMP\_F} \cdot$$

Now, the basic resolution definitions, which have been stated in Chapter 3.3.1, must be taken into consideration. Thus, for a total resolution A_{Res} it is valid:

$$A_{Res} = m + n \ , \tag{4.11}$$

and

$$n = n_1 + n_2, \text{ where } n_1, n_2 \in \mathbb{N},\ 1 \le n_1 \le n \text{ and } 0 \le n_2 \le n - 1 \ . \tag{4.12}$$

We can see that the first three terms in Eq. (4.10), containing S_{Ref}, S_{FA}, and S_{A_ITPL}, are strongly influenced by the interpolating resolution n_2. Providing the interpolating resolution n_2 is increased, the first two terms in Eq. (4.10) will decrease, but the 3rd term increases exponentially. The last two terms, containing S_{CMP_C} and S_{CMP_F}, in Eq. (4.10) are independent of the interpolating resolution n_2 and depend only on the coarse resolution and the fine resolution.

When considering Eqs. (4.10)~(4.12), it becomes obvious that a fundamental optimization without knowing the real values of S_{Ref}, S_{FA}, S_{A_ITPL}, S_{CMP_C}, and S_{CMP_F} of the components is impossible.

4.2.3 Power Dissipation Analysis

Considering the estimate of the power dissipation, we can basically use Eq. (4.8) and replace the chip area size S by power dissipation P. Since it can be argued that circuits exhibiting large chip area also tend to consume more power, Eq. (4.8) remains basically valid. Nevertheless, this is not necessarily true for resistors: large resistors tend to dissipate less power and this must be accounted for. As a result, power dissipation in resistors must be considered carefully.

The 1st term in Eq. (4.8) is related to the resistor area, which corresponds to the resistors which generate reference voltages. In order to meet the requirement of resolution, the power dissipation is proportional to the number of representable states. The power dissipation of this part can be written as:

$$2^{n_1} \cdot (2^m + 2) \cdot P_{Ref} \,. \tag{4.13}$$

The 2nd term includes now only the contribution of the folding amplifiers similarly as in Eq. (4.4) and but without the contribution of resistors included in Eq. (4.5). Thus, the power dissipation of the folders can be described as:

$$2^{n_1} \cdot (2^m + 3) \cdot P_{FA} \,. \tag{4.14}$$

The 3rd term in Eq. (4.8) represents the die area of interpolation resistors, which dissipate power supplied from the fourth term in Eq. (4.8). Hence, for a more exact estimate of power dissipation, the third term in Eq. (4.8) is eliminated. The scaling factor of the interpolation stage for the power estimate is only P_{A_ITPL}, instead of P_{R_ITPL} and P_{A_ITPL}. The last two terms in Eq. (4.8) stay unchanged.

The formula of power dissipation estimate can be then summarized as:

$$\begin{aligned} &2^{n_1} \cdot \left[(2^m + 2) \cdot P_{Ref} + (2^m + 3) \cdot P_{FA} + (2^{n_2} - 2) \cdot P_{A\_ITPL} \right] + \\ &(2^m + 1) \cdot P_{CMP\_C} + 2^n \cdot P_{CMP\_F} \,. \end{aligned} \tag{4.15}$$

4.3 Design and Implementation of the 10-Bit Folding and Interpolating A/D Converter

In this chapter, we will describe design and implementation of a 10-bit folding and interpolating A/D converter. 10 bit resolution was chosen as typical requirement for today's high-speed converters.

4.3.1 Optimization of the System Resolution

It is surmised from Eqs. (4.10) and (4.15) that the best trade-off between the die area and the power dissipation cannot be achieved without knowing the

Table 4.1: Total number of comparators

Architectures	0/10*	1/9	2/8	3/7	4/6	5/5	6/4	7/3	8/2	9/1	10/0*
Coarse comparators	0	3	5	9	17	33	65	129	257	513	1025
Fine comparators	1024	512	256	128	64	32	16	8	4	2	0
Total	1024	515	261	137	81	65	81	137	261	515	1025

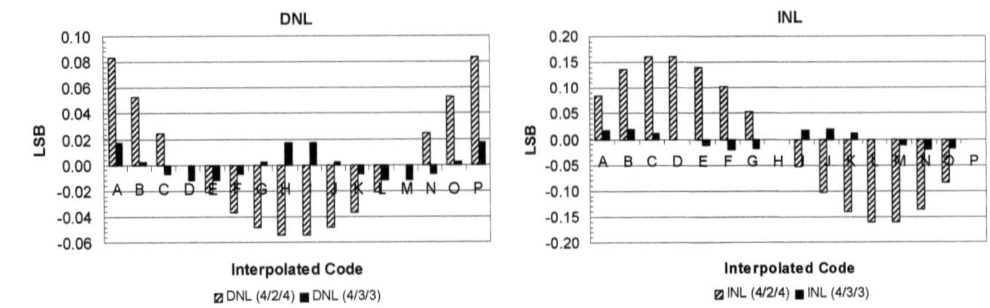

Fig. 4.2: DNL and INL caused by the interpolation error

exact values of all variables in those equations. On the other side, without knowing the best trade-off, the distribution of the system resolution between coarse and fine converters cannot be properly made.

To overcome this impasse, our solution is to assume a possible combination of the coarse and fine resolutions, implement this architecture and determine the real values of those variables in Eqs. (4.10) and (4.15) from the implementation. Then we feed these values back into the equations to predict performance of other combinations [48]. Iterating this process serves to build a *"learning curve"* which makes the prediction of A/D converter performance more exact.

Since the high-speed comparator plays an important role in the design, determination of the combination of the coarse and fine resolutions begins with determination of the total number of high-speed comparators needed. To simplify the calculation, comparators for the coarse and fine converters are considered to be the same. For a 10-bit folding and interpolating converter, the total number of comparators for different m/n resolution combinations is listed in Table 4.1.

First of all, Table 4.1 shows that the total number of comparators is symmetrical around the 5/5-architecture. Since the coarse comparator has to meet more stringent accuracy requirements than the fine comparator (see Chapter 2.5.6), only the first 6 columns are of interest. The 0/10- and 10/0-architectures are indeed flash A/D converters, thus, they are marked with an "*".

Second, the three "good" candidates, i.e. the combination of 3/7-, 4/6-, and 5/5-architectures, are marked in bold characters. Although the 5/5-architecture uses the lowest number of comparators than other architectures, it has more coarse comparators, which need more accuracy. Besides, considering the problem of internal frequency (see Chapter 2.5.6), which is proportional to the folding factor F_F, the 5/5-architecture is not the best choice because $F_F = 2^m$. Thus, the first implementation chosen here is based on the 4/6-architecture.

Further, the 6-bit fine resolution has to be split into the folding resolution n_1 and the interpolating resolution n_2. From Eqs. (4.10) and (4.15), we can see that if m and n are fixed, the folding resolution n_1 greatly affects both, die area and power dissipation. Thus it should be low as possible. This means that the interpolation resolution should be as high as possible. However, there is a limit on interpolation resolution: using the model of interpolation error discussed in Chapter 3.3.5 can help us to find the interpolation resolution.

The DNL and INL caused by the interpolation error between the neighboring folding signals by 4/2/4- and 4/3/3-architectures of the 10-bit folding and interpolating A/D converter with 16-times interpolation and 8-times interpolation, respectively, are illustrated in Fig. 4.2. This shows that the static performance of the 4/2/4-architecture A/D converter is much worse than the 4/3/3-architecture A/D converter. Besides, the amplitude distortion will increase when the number of the generated interpolation signals of single stage increases. The calculation assumes that the folding signal is sinewave.

Another possibility is to use the 4/4/2-architechture. As discussed in Chapter 3.4.1, the input signal range is divided into $(2^m \cdot 2^{n_1})$ intervals by the folding stage. If we consider the fact that the $0.6\ \mu m$ CMOS technology to be used for implementation allows a power supply of $3.3\ \text{V}$, the usable input signal range of the converter will be about half this voltage, i.e. ca. $1.6\ \text{V}$. If we also choose $V_{Ref} \cong 1.6\ \text{V}$ and estimate $\Delta R/R$ to lie in the range 0.5~1% [106], we obtain $\Delta V_{OS,\,Ref,\,\max} \cong 2\ \text{mV}$ for $\Delta R/R = 0.5\%$ (simplified Eq. (3.107)). If we further estimate that $\Delta V_{OS,\,CMP,\,\max}$ is going to be of the same order we get again about $2\ \text{mV}$. The noise of the resistor string can be calculated for estimated values $2^{m+n_1} = 256$ and $C_{In} = 2.5\ \text{pF}$ as $\sqrt{\overline{v_{n,\,Ref,\,\max}^2}} = 620\ \mu\text{V}_{\text{RMS}}$ (see Eq. (3.111)). Finally, the comparator noise can be guessed using Eq. (3.119) (and assuming e.g. $k' = 2 \times 10^{-24}\ \text{V}^2\text{F}$, $C_{OX} = 1.77\ \text{fF}/\mu\text{m}^2$, $C_{Out} = 100 \cdot C_A$, $W_1 = 100\ \mu\text{m}$, $W_3 = 3\ \mu\text{m}$, $L_1 = L_3 = 0.6\ \mu\text{m}$, $g_{m1}/g_{m3} = 10$, $g_{m3} \cong 0.1 \cdot (g_{m7} + g_{m9})$, and $g_{m5} \gg g_{m7} + g_{m9})$ as $\overline{v_{n,\,CMP}^2} = (190\ \mu\text{V}_{\text{RMS}})^2$. The resulting error voltage (Eq. (3.122)) is about $5.7\ \text{mV}$. Hence, the voltage drop between two resistors which $1.6\ \text{V}/2^{4+4}$ for

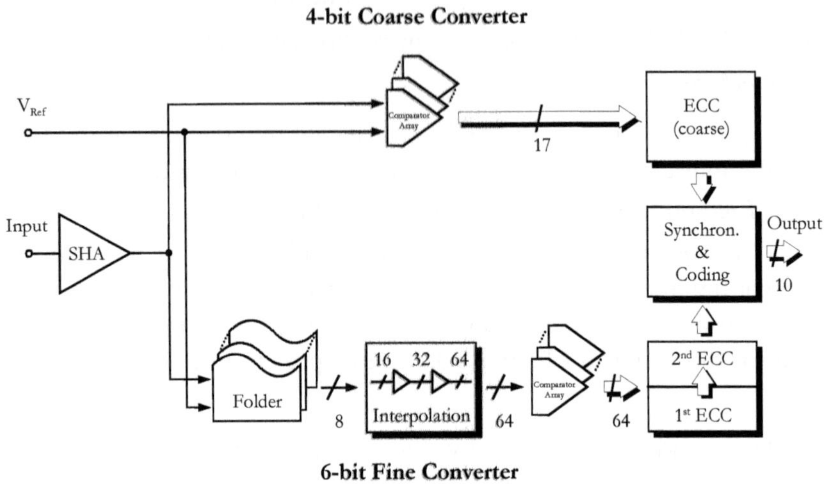

Fig. 4.3: Block diagram of the investigated folding and interpolating A/D converter

4/4/2-architecture is too close to the error voltage and this architecture is not suitable.

According to the facts discussed above, the 4/3/3-architecture is the best candidate for the implementation.

4.3.2 System Description

A block diagram of the implemented folding and interpolating A/D converter is depicted in Fig. 4.3 which is identical with Fig. 2.19, but it includes the resolution in bits. The input signal is sampled by the double-sampled THA and the output of the double-sampled THA is simultaneously fed into comparators of the coarse converter and folders of the fine converter. Since the double-sampled THA and the comparators of the coarse converter are controlled by the system clock and output signals of comparators are latched, problems of signal racing can be prevented. Since the chosen architecture is 4/6, the coarse converter has resolution of 4-bit. Including the out-of-range indicator, there are total 17 comparators in the coarse converter.

Both the folding resolution and the interpolating resolution are of 3-bit, i.e. the folding and interpolating A/D converter is built in 4/3/3-architecture. Together they yield a fine converter of 6-bit resolution. As a result, there are 8 folders connected in parallel and a cascaded 3-stage 2-times interpolation,

Fig. 4.4: Photomicrograph of the implemented 10-bit folding and interpolating A/D converter

which generates an interpolating factor of 8. The total number of output signals of the interpolation stage is 64 and the output signals of the interpolation are fed into comparators, which are controlled by the system clock, of the same number.

The general expression for the die area of the implemented folding and interpolating A/D converter, given by Eq. (4.10), can be specified as:

$$\begin{aligned} & 8\cdot\begin{bmatrix}(16+2)\cdot S_{Ref}+(16+4)\cdot S_{FA}\\ +(1.8\cdot 8-2.8)\cdot S_{A\_ITPL}\end{bmatrix}+(16+1)\cdot S_{CMP\_C}+64\cdot S_{CMP\_F} \\ & =144\cdot S_{Ref}+160\cdot S_{FA}+92.8\cdot S_{A\_ITPL}+17\cdot S_{CMP\_C}+64\cdot S_{CMP\_F}\,. \end{aligned} \quad (4.16)$$

In contrast, the power dissipation of the implemented folding and interpolating A/D converter, based on Eq. (4.15), can be expressed as:

$$\begin{aligned} & 8\cdot\begin{bmatrix}(16+2)\cdot P_{Ref}+(16+3)\cdot P_{FA}\\ +(8-2)\cdot P_{A\_ITPL}\end{bmatrix}+(16+1)\cdot P_{CMP\_C}+64\cdot P_{CMP\_F} \\ & =144\cdot P_{Ref}+152\cdot P_{FA}+48\cdot P_{A\_ITPL}+17\cdot P_{CMP\_C}+64\cdot P_{CMP\_F}\,. \end{aligned} \quad (4.17)$$

Since the digital error correction stages are directly connected to the output of the coarse and fine comparators, the die area and the power dissipation of the digital error correction are related to the total number of its input signals, which is related to the number of comparators that used as it can be seen in Fig. 4.3. Hence, the scaling of the digital part is correlated with the total number of comparators in the corresponding comparator stage.

A photomicrograph of the implemented chip is illustrated in Fig. 4.4. This chip has been fabricated in the 0.6 μm standard double-well CMOS technology (C0512) of Fraunhofer Institute of Microelectronic Circuits and Systems in Duisburg. This smallest channel length of this technology is 0.6 μm and three metal layer and one low-ohmic polysilicon layer (3M1P) are available in this technology. A high-ohmic polysilicon layer was added for realization of the resistor string. The width and height of the core are 3.050 mm and 1.250 mm, respectively, and the core area amounts to 3.8125 mm^2. While the sampling rate is at 40 MHz, the analog part dissipates 20 mA at 3.3 V power supply voltage and the digital part dissipates 12 mA (also at 3.3 V). When the sampling rate is at 20 MHz, the analog part dissipates 19 mA and the digital part dissipates 7 mA, all being operated at power supply voltage of 3.3 V.

4.3.2.1 Track-and-Hold Amplifier

The track-and-hold amplifier used here is the double-sampled THA shown in Fig. 3.3(d). As mentioned in Chapter 3.1, its core component is the high-speed OTA analyzed in Chapter 3.2. We have designed the OTA for requirements given by our 10-bit folding and interpolating A/D converter.

The OTA we employed in the double-sampled THA is that of Fig. 3.8 which features folded-cascode and gain-boosting. The bandwidth calculation was carried out for 20 and 40 MHz clock operation.

In case that both the rise time t_R and the fall time t_F are 0.5 $n$sec and the time gap t_{NP} of the non-overlapped clock signals is 2 $n$sec, the pulse width t_H of the clock signal is only 9.5 $n$sec for 40 MHz operation.

Hence, for an accuracy of 10-bit and the pulse width t_H of 9.5 $n$sec, the required bandwidth is (assuming $G_m R_{Out} > 1000$, $G_m = g_{m1} = g_{m2}$, $V_{OS} \cong 10$ mV, $C_L = 1.5$ pF, $V_{In,\,\max} = 1.6$ V, $V_{In,\,\min} = V_{In,\,\max} / 2^{m+n} = V_{In,\,\max} / 2^{10}$):

$$f_{3dB,\,Hold} = 134 \text{ MHz}. \tag{4.18}$$

If the pulse width t_H is extended to 22 $n$sec as for 20 MHz, the estimated bandwidth can be written as:

$$f_{3dB,\,Hold} = 58\ \text{MHz}\,. \tag{4.19}$$

The transconductance $G_m = g_{m1} = g_{m2}$ can be then determined as:

$$G_m = 2\pi \cdot C_L \cdot f_{3dB,\,Hold}\,, \tag{4.20}$$

and yields $1.26\ \text{mS}$ for $40\ \text{MHz}$ and $0.55\ \text{mS}$ for $20\ \text{MHz}$, respectively. Note that the effect C_M has been neglected as C_M can be chosen much smaller than C_L.

The noise of the OTA can be then estimated as follows. For a properly designed OTA is the input transistor pair M1 and M2 which determines the transconductance. Then the equivalent input voltage

$$\overline{v_n^2} \cong \frac{8}{3} \cdot \frac{kT}{C_L} = \left(86\ \mu\text{V}_{RMS}\right)^2 \quad \text{for } C_L = 1.5\ \text{pF}. \tag{4.21}$$

Considering the range of $1.6\ \text{V}$, the noise is 2.6 bit below the required resolution. Hence only the effect of the deterministic error had to be considered (Eqs. (4.18) and (4.19)).

4.3.2.2 Reference Voltages

The reference voltages used in the coarse converter can be calculated using the following relation:

$$\Delta V_{Ref\_C} = \frac{V_{In\_\max} - V_{In\_\min}}{2^m} = \left(\frac{2.25 - 0.65}{2^4}\right) \text{V} = 100\ \text{mV}\,, \tag{4.22}$$

where the input range is given by $V_{In\_\max} = 2.25\ \text{V}$ and $V_{In\_\min} = 0.65\ \text{V}$. The limit is set by the output signal range of the double-sampled THA described in Fig. 3.3(d).

The voltage difference between neighboring reference voltages in the fine converter ΔV_{Ref_F} is much smaller than the voltage difference ΔV_{Ref_C} in the coarse converter. The variable ΔV_{Ref_F} depends only on the input voltage range and the interpolating factor F_{ITPL} and can be calculated as:

$$\Delta V_{Ref\_F} = \frac{V_{In\_\max} - V_{In\_\min}}{2^{m+n}} \cdot 2^{n_2} = \left(\frac{2.25 - 0.65}{2^{10}} \cdot 2^3\right) \text{V=12.5 mV}\,, \tag{4.23}$$

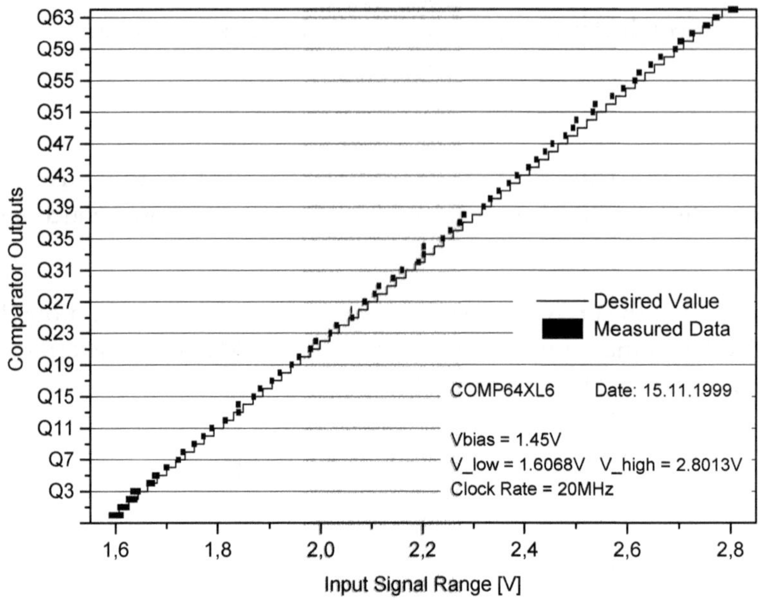

Fig. 4.5: Transfer function of a 6-bit flash A/D converter

which is already quite small compared to the error voltage 5.7 mV calculated according to Eq. (3.122) in Chapter 4.3.1. Assuming that the interpolating factor F_{ITPL} is 4, instead of 8, the variable ΔV_{Ref_F} will be halved, which is already too close to the error voltage. Thus, interpolating resolution of 3-bit is necessary. This is a heavy constraint for the architecture selection, since the more the signals are interpolated, the more interpolation errors are generated, but the die area and power dissipation are decreased.

Due to the fact that the reference voltages in the coarse converter are apart from each other far away (100 mV as given by Eq. (4.22)), the coarse converter needs only single stage for the error correction. The fine converter, in contrast to the coarse converter, has to have two stages for error correction to correct the potential errors in the fine thermometer codes. The output signals of the conversion system are coded in 2's complement code.

4.3.2.3 High-Speed Comparator

To investigate the behavior of the high-speed comparator proposed in Chapter 3.5.2, a prototype of a 6-bit flash A/D converter was fabricated at first by

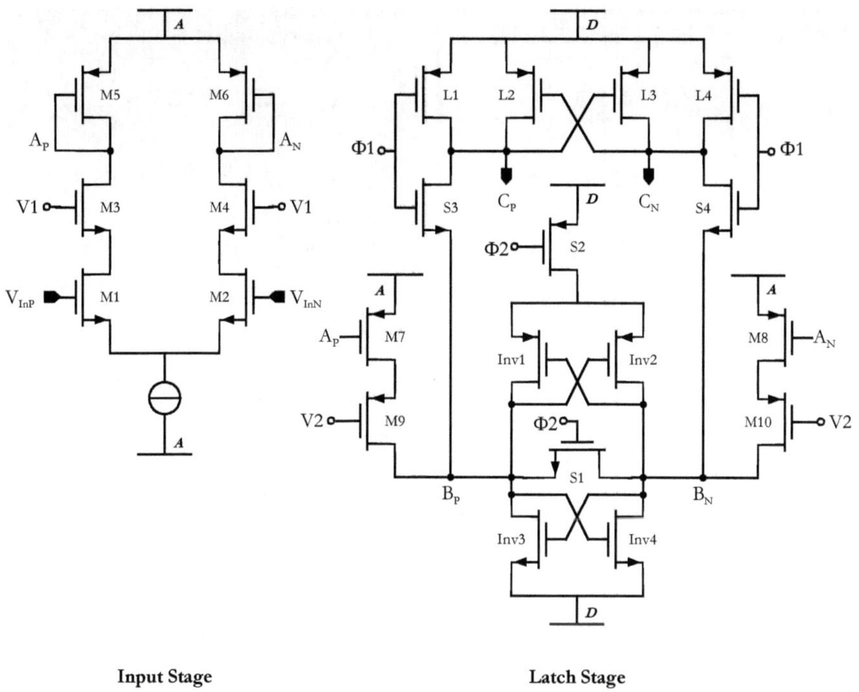

Fig. 4.6: Schematic of the clock crosstalk-improved comparator

stacking 64 high-speed comparators shown in Fig. 3.36. The A/D converter is based on the block diagram depicted in Fig. 2.10. The output signals of this flash A/D converter are kept at their original thermometer code, instead of coding them into binary code. This enables direct observation of the real behavior of the output signals of the stacked comparators.

The measurements of this 6-bit flash A/D converter (see Fig. 4.5) show that the thermometer codes of the stacked comparators are accompanied with numerous errors, which can be corrected partly by the concept of the digital error correction, but it turned out that for the sake of accuracy, more improvements must be undertaken.

The analysis of the measurements has yielded that the input signal suffers from the clock crosstalk, as pointed in Chapter 3.5.3. To prevent this an improved comparator has been developed that can be found in Fig. 4.6.

In contrast to the comparator shown in Fig. 3.34, the comparator circuit illustrated in Fig. 4.6 uses two cascode circuits (transistors M3, M4 and M9, M10), which isolate the input devices (transistors M1 and M2) from the load

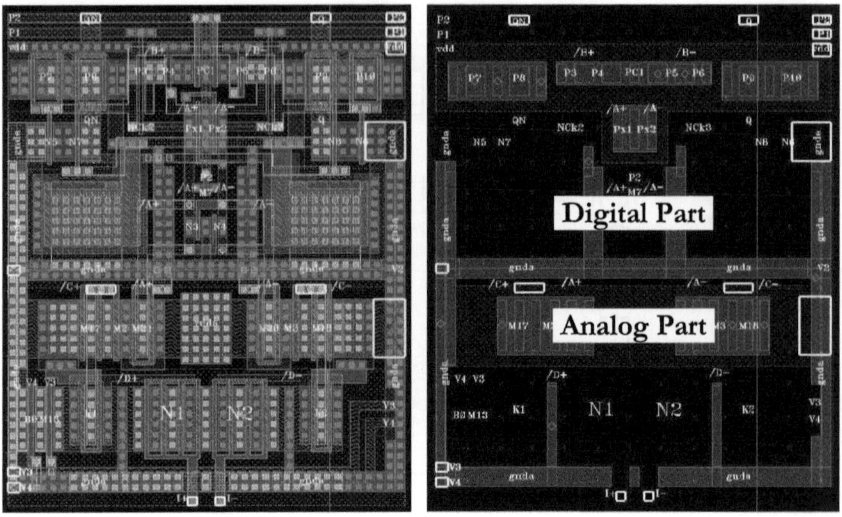

(a) (b)

Fig. 4.7: Layout of the clock crosstalk-improved comparator

devices (transistors M5 and M6) and latch drivers (transistors M7 and M8) from the latch and its switches. Apart from these additions, the analog and digital power supplies power supply voltages for the input stage and latch, respectively, are separated, in order to prevent the inherent crosstalk between the analog circuits and the digital parts of the comparator.

As indicated in the schematic, the nodes B_P and B_N are especially critical. These nodes are connected to the switches S_1, S_3, and S_4, which are controlled by the clock signals Φ_1 and Φ_2. The clock signals Φ_1 and Φ_2 used here are non-overlapped clock signals. The PMOS cascode transistors M9 and M10 can suppress the clock feedthrough more efficiently.

Furthermore, the layout of the comparator must be carried out very carefully. First of all, it is of great advantage to use the fact that the clock feedthrough in differential-path circuits may cancel if it is applied to both paths in the same way. Thus, layout symmetry for such circuits is a must so that all parasitic capacitances are also fully symmetrical, as illustrated in Fig. 4.7. Paired transistors in the 1st fabricated comparator (Fig. 3.36) were designed with *local layout symmetry*. The transistor pairs in the 2nd test chip of comparators shown in Fig. 4.7(a), however, have been designed to exhibit local layout symmetry as well as through the entire layout of the comparator (*global layout symmetry*) to keep the circuit "balanced".

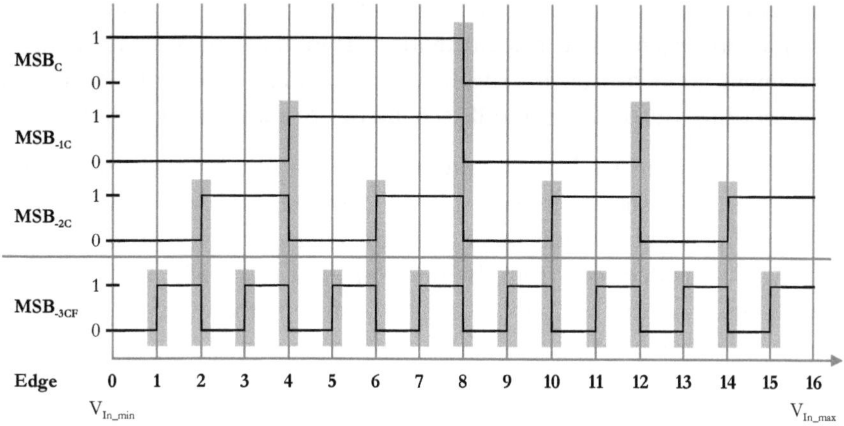

Fig. 4.8: Synchronization scheme for the coarse A/D converter

Secondly, as illustrated in Fig. 4.7(b), the new layout separates the analog and digital power supplies already in the comparator stage. The well contacts have also been separated into two groups: one is for the analog power supply voltage and another for the digital power supply voltage. This arrangement also places the digital PMOS transistors of all comparators on the same row in the same well (NSUBs, printed in red in Fig. 4.7(b)) to reduce the potential difference from the different well voltages and to drain away the digital switching currents as soon as possibly by attracting carriers injected due to the switching transients.

Another measure to reduce the clock feedthrough in this stage is to introduce the averaging technique [71]. The averaging resistors are connected at both sides of the comparator inputs.

4.3.2.4 Realization of the Synchronization

For the implemented 10-bit folding and interpolation A/D converter, the resolutions of the coarse and fine converters are 4-bit and 6-bit, respectively. Hence, the *LSB* of the coarse converter is 6th bit of the whole converter. It can be also named as MSB_{-3} to indicate the relation to the *MSB* of the whole converter.

Since the MSB_{-3} in the coarse converter is deduced from the thermometer code of the fine converter, it offers possibility for synchronization between the two converters. A detailed description about the synchronization will be handled in the later section of this chapter.

The final binary coding is a 2's complement coding for the following digital signal processing. The transformation from the 1-of-N code to the 2's complement code has been executed using a CAD-software Verilog. The corresponding source codes and its schematic are shown in Appendix B.

A complete synchronization scheme for a 4/6 folding and interpolation A/D converter is shown in Fig. 4.8. All three signals MSB_C, MSB_{-1C}, and MSB_{-2C} in Fig. 4.8 are signals taken from the coarse A/D converter. Since the signal MSB_{-3CF} is deduced from the fine converter (similarly as in Table 3.4), it can be used as the basis for synchronization of the other signals in the coarse converter with the signals from the fine converter.

The edges in Fig. 4.8 can be sorted into three groups. The first edge group includes edges 2, 6, 10, and 14. The second edge group includes edges 4 and 12. The third edge group includes the edge 8 alone. All three edge groups will be discussed separately.

1st edge group: edges 2, 6, 10, and 14

Detection logic:

$$(Critical\_Region) \wedge (Edge_2 \vee Edge_6 \vee Edge_{10} \vee Edge_{14})$$
$$= (Critical\_Region) \wedge \begin{Bmatrix} [(Coarse\_TCode_1) \wedge (Coarse\_NTCode_3)] \vee \\ [(Coarse\_TCode_5) \wedge (Coarse\_NTCode_7)] \vee \\ [(Coarse\_TCode_9) \wedge (Coarse\_NTCode_{11})] \vee \\ [(Coarse\_TCode_{13}) \wedge (Coarse\_NTCode_{15})] \end{Bmatrix} . \quad (4.24)$$

Synchronized signals in the coarse converter:

$$\begin{aligned} MSB &= MSB_C \ , \\ MSB_{-1} &= MSB_{-1C} \ , \\ MSB_{-2} &= \text{not}\,(MSB_{-3CF}) \ . \end{aligned} \quad (4.25)$$

2nd edge group: edges 4 and 12

Detection logic:

$$(Critical\_Region) \wedge (Edge_4 \vee Edge_{12})$$
$$= (Critical\_Region) \wedge \begin{Bmatrix} [(Coarse\_TCode_3) \wedge (Coarse\_NTCode_5)] \vee \\ [(Coarse\_TCode_{11}) \wedge (Coarse\_NTCode_{13})] \end{Bmatrix} . \quad (4.26)$$

Synchronized signals in the coarse converter:

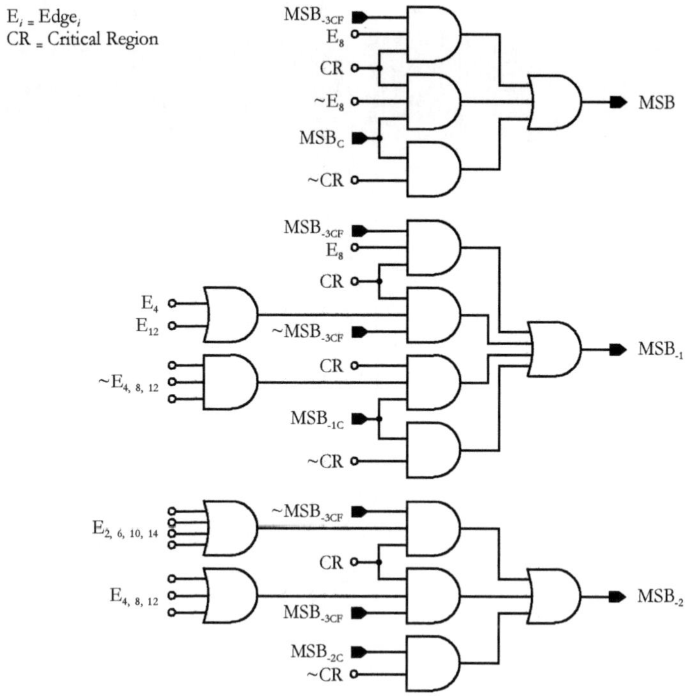

Fig. 4.9: Logic for the output synchronization

$$\begin{aligned} MSB &= MSB_C \ , \\ MSB_{-1} &= \text{not}\,(MSB_{-3CF}) \ , \\ MSB_{-2} &= MSB_{-3CF} \ . \end{aligned} \tag{4.27}$$

3rd edge group: edge 8

Detection logic:

$$\begin{aligned} &(Critical\_Region) \wedge (Edge_8) \\ &= (Critical\_Region) \wedge \{(Coarse\_TCode_7) \wedge (Coarse\_NTCode_9)\} \ . \end{aligned} \tag{4.28}$$

Synchronized signals in the coarse converter:

$$\begin{aligned} MSB &= MSB_{-3CF} \ , \\ MSB_{-1} &= MSB_{-3CF} \ , \\ MSB_{-2} &= MSB_{-3CF} \ . \end{aligned} \tag{4.29}$$

Fig. 4.10: Comparators array inside the fine converter

The synchronization logic schematic that has been derived from the equations above is shown in Fig. 4.9. By using some logic gates, the output signals of the coarse A/D converter MSB_{C}, MSB_{-1C}, and MSB_{-2C} are synchronized with the output signal MSB_{-3CF}, which is generated from the thermometer codes of the fine A/D converter as discussed above.

4.4 Architectural Scalability of the Folding and Interpolating A/D Converter

4.4.1 Introduction

Aim of this section is to define the rating scale for the architectural scaling of the folding and interpolating A/D converter. The rating scale will be calculated both for the optimization of the power dissipation and the die area. To

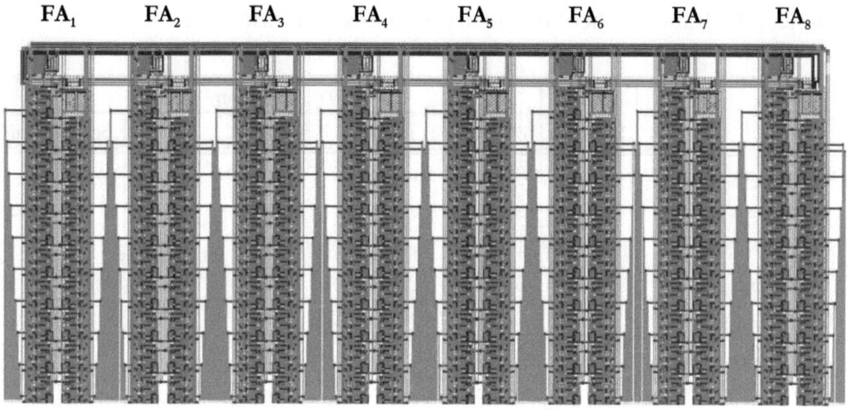

Fig. 4.11: Stacked folding blocks in the folding stage

define and calculate the rating scale, some design aspects must be considered.

First of all, we have to consider the die circuit area and power dissipation. Models of the die area estimate and the power dissipation estimate have been already derived in Chapters 4.2.2 and 4.2.3. In the following we are going to consider converter scaling with respect to die area and power dissipation based on these models. The estimates of die area and power dissipation consumption will be considered separately.

Secondly we have to consider connecting wires. The connecting wires play an important role when considering the die area, but it is very difficult to estimate the chip area required because of the tremendous number and the complexity of the connecting wires. A reasonable trade-off to solve this problem is to add the area of the connecting wires to the circuit area together.

When carrying the geometrical layout, the basic components of the folding and interpolating A/D converter can be stacked in a very regular manner. These basic components are going to be considered as inseparable modules, e.g. a single fine comparator will be treated as the module for the comparator array and the folding block will be considered as the module for the folding stage. Layout examples of the fine comparator array and the stacked folding blocks inside the folding stage are shown in Fig. 4.10 and Fig. 4.11, respectively.

Table 4.2: Balance sheet of die areas in 4/3/3-architecture

	# of	Die Area (mm²)				
	Modules	W (mm)	H (mm)	Area	%	Scaling Unit
Double-Sampling SHA	1	0.750	0.275	0.206250	6.3%	0.206250
Reference Voltages	144	1.155	0.235	0.271425	8.3%	0.001885
Folding Stage	160	1.130	0.710	0.802300	24.5%	0.005014
Interpolating Stage	87.648	0.670	0.250	0.167500	5.1%	0.001911
Coarse Comparators	17	0.235	1.000	0.235000	7.2%	0.013824
Fine Comparators	64	0.845	0.785	0.663325	20.3%	0.010364
Coarse ECC	17	0.225	0.470	0.105750	3.2%	0.006221
Fine ECC	64	0.885	0.840	0.743400	22.7%	0.011616
Synchronization	1	0.200	0.390	0.078000	2.4%	0.078000
Sum				3.272950	100.0%	
Core				3.812500		

4.4.2 Scaling Unit for Die Area

Based on the layout of the 10-bit folding and interpolating A/D converter implemented in CMOS technology and based on 4/3/3-architecture, basic components used in building block have been characterized and these basic components are used as modules for performance estimate of other architectures. A balance sheet for the die areas of different building blocks accompanied with the total number of needed modules in each building block is listed in Table 4.2. The die areas are valid for 0.6 μm standard double-well CMOS technology with 3 metal layers and 1 polysilicon layer (C0512) available at the Fraunhofer Institute of Microelectronic Circuits and Systems in Duisburg.

The double-sampled THA and the synchronization stage in Table 4.2 are not scalable and must be considered as a whole component with a fixed die area. Furthermore, the total number of modules for reference voltages, folding stage, interpolation stage, and comparator stages in Table 4.2 are derived from Eq. (4.16). The digital error corrections and coding (ECC) of the coarse and fine converters in Table 4.2 are matched to the comparator stages.

It can be also observed from Table 4.2 that the folding stage, the comparator stage, and the two error correction stages of the fine converter occupy 67.5% of the total chip area. Providing that a small die area of the complete conversion system is of primary importance for the application, area reduc-

Table 4.3: Balance sheet of power dissipation (analog part)

	# of Modules	Power (mW)		
		Power	%	Scaling Unit
Double-Sampling SHA	1	18.3	32.7%	18.3000
Reference Voltages	144	3.0	5.4%	0.0208
Folding Stage	160	19.1	34.1%	0.1194
Interpolating Stage	48	4.2	7.5%	0.0875
Coarse Comparators	17	2.4	4.3%	0.1412
Fine Comparators	64	9.0	16.1%	0.1406
Coarse ECC	17	–	–	–
Fine ECC	64	–	–	–
Synchronization	1	–	–	–
Sum		56	100.0%	

tion effort should concentrate on these parts. On the right side of Table 4.2, die areas of each module including wiring are given for further estimates. The wiring area is contained within each building block and assumed that the wiring area is proportional to the number of modules.

4.4.3 Scaling Unit for Power Dissipation

Modeling of power dissipation is also based on the simulation results of the implemented 10-bit folding and interpolating A/D converter. A balance sheet of power dissipation of this A/D converter is listed in Table 4.3. The simulation has been carried out using the tool "Analog Artist" from Cadence and provides only results for the analog circuits. The simulated total power dissipation of the analog part is equal to $56\ \mathrm{mW}$, which is very close to the measured result of $59.4\ \mathrm{mW}$, as discussed in the following chapter.

The digital part, including the error correction and coding stages (ECC) of the coarse and the fine converter and the synchronization stage, has been simulated using another tool called "Verilog", which does not allow an estimate of power dissipation. The power dissipation of the digital circuits can be measured at realized chips because the digital power supply voltage is separated from the analog power supply voltage during the measurement; this will be discussed in Chapter 5. Because most of the digital power dissipation is caused by the output pad drivers of the prototype, the power dissipation will be much lower in embedded applications, where such drivers are not necessary. Hence, in the following we concentrate on estimate of the analog part dissipation.

It can be observed from Table 4.3 that double-sampled THA (DS-THA) stage needs 32.7% of total analog power dissipation, although the THA stage used here dissipates only $18.3\ \mathrm{mW}$. This is already quite low when compared to $70\ \mathrm{mW}$ at $5\ \mathrm{V}$ of another implementation [71, 72].

Let us assume that the A/D converter operates under the condition of oversampling, that is to say, the input signal frequency f_{In} is much lower than the clock frequency f_{Clk}. In such a case, the internal frequency inside the folding stage is also low. As a result, the DS-THA stage can be removed in order to reduce the power dissipation. According to the simulation results of Table 4.3, it takes $37.7\ \mathrm{mW}$ for the analog core.

Another possibility to reduce the power dissipation is when the folding and interpolating A/D converter is to be embedded into systems, which have already THA as their output signal buffer. As an example, imaging sensors usually have an integrated THA working as an output buffer. Thus, an extra THA at the front-end of the A/D converter is redundant.

4.4.4 Scaling Unit for Input Capacitance of Interstages

4.4.4.1 Low-Power Design

Generally speaking, there are three different ways to interconnect building blocks. One is the serial connection, the second one is the parallel connection, and the last possibility is combination of the first two methods. Cascading building blocks, for example, such as multi-stage folding or multi-stage interpolation, increases group delay time and thus causes longer latency of the whole conversion system. As it can be seen in Fig. 4.3, only multi-stage interpolation has been implemented the 10-bit folding and interpolating A/D converter presented in this book. Since the multi-stage interpolation contains single-stage amplifiers and interpolating resistors, as shown in Fig. 4.1, the latency problem is not critical as that of multi-stage folding. The constraint on conversion rate is given by the fact that the total latency of the fine converter cannot exceed the reciprocal value of the sampling rate; otherwise, the fine comparator cannot be synchronized with the DS-THA stage.

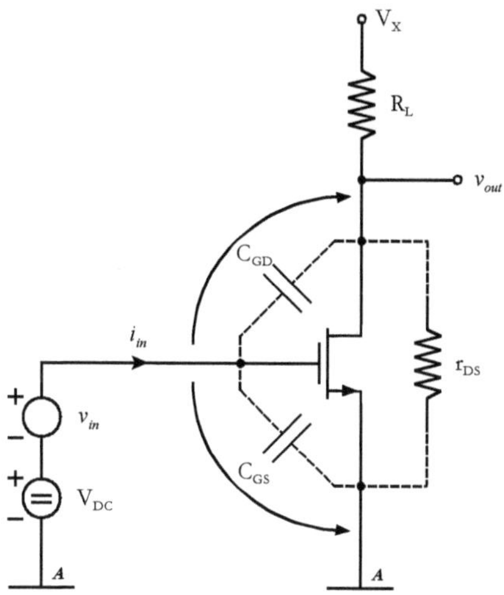

Fig. 4.12: Miller effect due to the gate capacitance

Since the input impedances of the building blocks basically exhibit capacitive behavior, connecting building blocks in parallel tends to raise the input impedance, which can exceed what the previous stage can drive. Parallel stacking of the components can be found in the folding stage, the interpolation stage, and the comparator stage and also in the digital error correction stage. Hence, investigation concerning the input impedance of each building block is essential for modeling of a single building block. The input impedance affects the power dissipation as it represents load for preceding building blocks. Optimizing the input impedance can thus help to reduce power dissipation.

When considering the CMOS realization, except for the interpolation stage, the input stages of all other building blocks of the A/D converter contain MOS transistors with their gates forming the input terminals. The gate capacitance is defined by the gate-drain capacitance C_{GD} and the gate-source capacitance C_{GS} as shown in Fig. 4.12.

An input impedance of a simple MOS transistor in common source configuration is

$$\frac{1}{Z_{In}(s)} = \frac{i_{in}(s)}{v_{in}(s)} = sG_{GS} + \frac{sC_{GD}\cdot(1+|A_{V0}|)}{1+sC_{GD}\cdot(R_L \,||\, r_{DS})}, \tag{4.30}$$

where $A_{v0} = -g_m \cdot (R_L \| r_{DS})$ and $s = j\omega$. The voltage gain is

$$A_v(s) = \frac{v_{out}(s)}{v_{in}(s)} = A_{v0} \cdot \frac{1-(sC_{GD})/g_m}{1+sC_{GD} \cdot (R_L \| r_{DS})} . \tag{4.31}$$

Hence, the input impedance is clearly affected by both, C_{GD} and C_{GS}. Note that C_{GD} "appears" at the input terminal multiplied by the low-frequency voltage gain: this is the well-known Miller effect. For linear small-signal operation of MOS transistors in triode region, these capacitances can be approximated to the first order by [107]:

$$C_{GS} \cong C_{GD} \cong \frac{C_{OX} \cdot W \cdot L}{2} , \tag{4.32}$$

where W and L are the gate width and length, respectively, and C_{OX} is the *gate oxide capacitance per unit area.*

For linear small-signal operation in the saturation region, these capacitances can be expressed as:

$$C_{GS} = \frac{2 \cdot C_{OX} \cdot W \cdot L}{3} , \tag{4.33}$$

and

$$C_{GD} \cong 0 . \tag{4.34}$$

When evaluating simulations, the input capacitance C_{In} can be easily found as:

$$\begin{aligned} C_{In} &= \mathrm{Im}\left[\frac{1}{\omega \cdot Z_{In}(j\omega)}\right] = \mathrm{Im}\left[\frac{i_{in}(j\omega)}{\omega \cdot v_{in}(j\omega)}\right] \\ &= f(\mathit{operating\ point}) . \end{aligned} \tag{4.35}$$

C_{In} is a function of operating point because the transconductance g_m itself is a function of operating point.

4.4.4.2 Computation of Input Capacitance

The input capacitance C_{In} of each building block of the 10-bit folding and interpolating A/D converter can be determined using AC simulations at different operating points. The largest input capacitance $C_{In,max}$ has been chosen as the worst case of each stage.

Table 4.4: Estimated input capacitance C_{In} of the THA stage

$V_{In,DC}$=0.65 [V]			$V_{In,DC}$=1.45 [V]			$V_{In,DC}$=2.25 [V]		
f [Hz]	i [A] in dB	C_{In} [F]	f [Hz]	i [A] in dB	C_{In} [F]	f [Hz]	i [A] in dB	C_{In} [F]
1.0E+03	148.76	5.80E-12	1.0E+03	148.62	5.90E-12	1.0E+03	148.64	5.89E-12
1.0E+06	91.93	4.03E-12	1.0E+06	91.76	4.11E-12	1.0E+06	91.81	4.09E-12

Sample-and-Hold Amplifier Stage

The designated input voltage range of the THA stage is between $0.65\ \text{V}$ and $2.25\ \text{V}$, which is also the voltage range of the operating points. The module THA has been simulated at three input DC voltages, namely at the minimum input voltage $0.65\ \text{V}$, the middle input voltage $1.45\ \text{V}$, and the maximum input voltage $2.25\ \text{V}$.

Table 4.4 summarizes the simulation results of the input capacitance C_{In} of the THA stage. The largest input capacitance $C_{In,max}$ of this stage is $5.9\ \text{pF}$ at $1\ \text{kHz}$. This value is very close to the hold capacitor C_H designated to $5\ \text{pF}$.

Table 4.5: Estimated input capacitance C_{In} of the comparator array

$V_{In,DC}$=0.65 [V]			$V_{In,DC}$=1.45 [V]			$V_{In,DC}$=2.25 [V]		
f [Hz]	i [A] in dB	C_{In} [F]	f [Hz]	i [A] in dB	C_{In} [F]	f [Hz]	i [A] in dB	C_{In} [F]
1.0E+03	173.00	3.56E-13	1.0E+03	172.20	3.91E-13	1.0E+03	171.10	4.43E-13
1.0E+06	113.01	3.56E-13	1.0E+06	112.20	3.91E-13	1.0E+06	111.15	4.41E-13

Comparator Array of the Coarse Converter

Table 4.5 shows the simulated input capacitance C_{In} of the comparator array of the coarse converter. The signals for the comparator array of the coarse converter and the folding stage of the fine converter are fed from the THA stage (Fig. 4.3). The largest input capacitance $C_{In,max}$ simulated for this comparator array is $4.43\ \text{fF}$ at $1\ \text{kHz}$. Note that there are 17 comparators in total connected in parallel in the array. Thus the input capacitance C_{In} of a single comparator is $0.26\ \text{fF}$.

Table 4.6: Estimated input capacitance C_{In} of the ECC of the coarse converter

V_{In}=0 [V]			V_{In}=1.65 [V]			V_{In}=3.3 [V]		
f [Hz]	i [A] in dB	C_{In} [F]	f [Hz]	i [A] in dB	C_{In} [F]	f [Hz]	i [A] in dB	C_{In} [F]
1.0E+03	159.61	1.66E-12	1.0E+03	155.19	2.77E-12	1.0E+03	162.39	1.21E-12
1.0E+06	99.70	1.65E-12	1.0E+06	95.19	2.77E-12	1.0E+06	102.51	1.19E-12

Digital Error Correction and Coding (ECC) of the Coarse Converter

The ECC stage differs since it contains digital circuits. Unlike analog circuits, digital circuits do not have DC operating point. Therefore, the ground, the mid point between the ground and the power supply voltage (i.e. at $3.3\ \text{V}/2 = 1.65\ \text{V}$), and the power supply voltage of $3.3\ \text{V}$ have been chosen as operating points for our simulation of input capacitance. The major input capacitance C_{In} amounts to $2.77\ \text{pF}$ (see Table 4.6) and occurs at the middle between the ground and the power supply voltage because that is the voltage where CMOS logic gates tend to switch their logical state and usually exhibit the maximum voltage gain. Hence, the input capacitance C_{In} is at its maximum $C_{In,\max}$ due to the Miller effect [79]. This digital ECC stage of the coarse converter has 17 voter circuits, as discussed in Chapter 3.6.2, connected in parallel at its input. As a consequence, the input capacitance of the single voter circuits can be estimated as $163\ \text{fF}$.

Table 4.7: Estimated input capacitance C_{In} of the folding stage

$V_{In,DC}$=0.65 [V]			$V_{In,DC}$=1.45 [V]			$V_{In,DC}$=2.25 [V]		
f [Hz]	i [A] in dB	C_{In} [F]	f [Hz]	i [A] in dB	C_{In} [F]	f [Hz]	i [A] in dB	C_{In} [F]
1.0E+03	164.10	9.93E-13	1.0E+03	164.90	9.05E-13	1.0E+03	166.50	7.53E-13
1.0E+06	104.33	9.67E-13	1.0E+06	105.00	8.95E-13	1.0E+06	106.50	7.53E-13

Folding Stage

According to Eq. (4.4) of Chapter 4.2.2.3, the total number of the scaling units, i.e. folding amplifiers, for the folding stage is 152, instead of 160 stated in Table 4.2 and Table 4.3. These tables considered not only the input circuits of the folding stage, but also the impact of the resistor load of the folders. The largest input capacitance $C_{In,\max}$ occurs at $1\ \text{kHz}$ and has amount of $993\ \text{fF}$ (see Table 4.7). The scaling unit of the folding stage can be estimated as $6.53\ \text{fF}$.

Table 4.8: Estimated input capacitance C_{In} of the interpolation stage

$V_{In, DC}$=0.65 [V]			$V_{In, DC}$=1.45 [V]			$V_{In, DC}$=2.25 [V]		
f [Hz]	*i* [A] in dB	C_{In} [F]	*f* [Hz]	*i* [A] in dB	C_{In} [F]	*f* [Hz]	*i* [A] in dB	C_{In} [F]
1.0E+03	165.12	8.82E-13	1.0E+03	170.09	4.98E-13	1.0E+03	171.01	4.48E-13
1.0E+06	105.13	8.82E-13	1.0E+06	110.09	4.98E-13	1.0E+06	111.01	4.48E-13

Interpolation stage

The largest input capacitance $C_{In,\max}$ occurs at low DC operating voltage and amounts to 882 fF (see Table 4.8). Providing that the operating voltage is shifted above the middle voltage 1.45 V the input capacitance C_{In} decreases to less than 500 fF. Since this interpolation stage is composed of three cascaded stages each performing 2-times interpolation (see Fig. 3.24 and Fig. 4.1), the input part of the interpolation stage relates has to carry out only first 2-times interpolation and has 16 distinguishable interpolating amplifiers. The input capacitance C_{In} of the scaling unit, namely the single interpolating amplifier, is then 31.25 fF.

Table 4.9: Estimated input capacitance C_{In} of the comparator array of the fine converter

$V_{In, DC}$=0.65 [V]			$V_{In, DC}$=1.45 [V]			$V_{In, DC}$=2.25 [V]		
f [Hz]	*i* [A] in dB	C_{In} [F]	*f* [Hz]	*i* [A] in dB	C_{In} [F]	*f* [Hz]	*i* [A] in dB	C_{In} [F]
1.0E+03	155.57	2.65E-12	1.0E+03	156.19	2.47E-12	1.0E+03	156.61	2.35E-12
1.0E+06	95.57	2.65E-12	1.0E+06	96.19	2.47E-12	1.0E+06	96.61	2.35E-12

Comparator Array of the Fine Converter

The comparator array of the fine converters contains 64 comparators connected in parallel. The largest input capacitance $C_{In,\max}$ occurs at the low DC operating voltage and has an amount of 2.65 pF. The input capacitance C_{In} of the single comparator is then 41 fF (see Table 4.9).

Table 4.10: Estimated input capacitance C_{In} of the ECC for the fine converter

V_{In}=0.3 [V]			V_{In}=1.65 [V]			V_{In}=3.3 [V]		
f [Hz]	*i* [A] in dB	C_{In} [F]	*f* [Hz]	*i* [A] in dB	C_{In} [F]	*f* [Hz]	*i* [A] in dB	C_{In} [F]
1.0E+03	140.63	1.48E-11	1.0E+03	138.49	1.89E-11	1.0E+03	142.71	1.16E-11
1.0E+06	80.66	1.47E-11	1.0E+06	78.48	1.90E-11	1.0E+06	82.76	1.16E-11

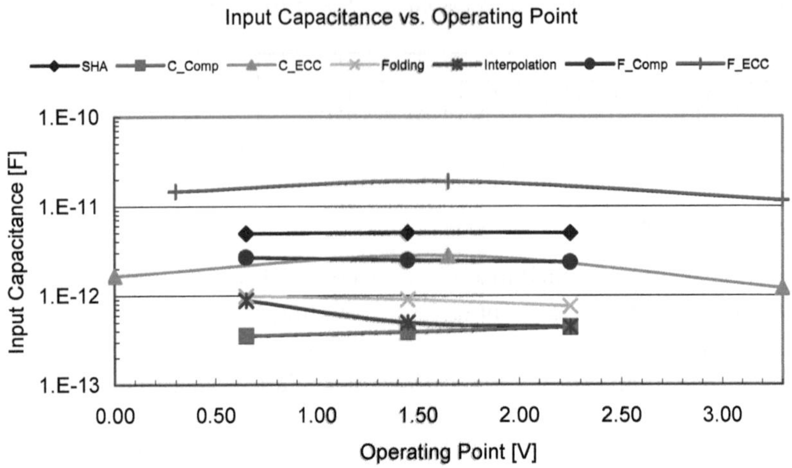

Fig. 4.13: Dependence of the input capacitance on the operating point

Digital Error Correction and Coding of the Fine Converter

Similarly as the error correction and coding of the coarse converter, the digital circuits here do not have DC operating points. Therefore, three points, i.e. 0.3 V, 1.65 V, and 3.3 V, have been chosen as operating points for our simulation of input capacitance (see Table 4.10). Due to the complexity of the convergence process of the simulation program, the lowest DC operating point of this stage can only reach 0.3 V, instead of 0 V.

The largest input capacitance $C_{In,max}$ of the digital ECC of the fine converter amounts to 19 pF and appears at the mid point between the ground and the power supply voltage, i.e. at 1.65 V. Since the ECC stage of the fine converter has 64 voter circuits at its input, the scaling factor of this stage is 64 and the input capacitance of each voter circuit is 297 fF.

Fig. 4.13 illustrates the variation of the input capacitance C_{In} in dependence of the operating points. Except for the interpolation stage, the input capacitances C_{In} of the analog circuits exhibit low dependence on DC operating voltages. The input capacitances C_{In} of the digital circuits reach the highest value at the mid point between the ground and the power supply voltage, since both the PMOS and the NMOS transistors are in saturation region and the voltage gain of logic gates is usually at its maximum.

Table 4.11: Estimate of the die area [mm²] as a function of parameter variations

	Coarse Resolution: m										
Folding Resolution: n_1	0	1	2	3	4	5	6	7	8	9	10
0	-	-	-	-	-	-	-	-	-	-	22.40
1	14.89	8.08	4.72	3.12	2.49	2.50	3.17	4.83	8.31	15.36	-
2	14.95	8.15	4.82	3.27	2.75	2.98	4.10	6.64	11.90	-	-
3	15.06	8.29	5.01	3.58	3.27*	3.95	5.94	10.28	-	-	-
4	15.28	8.57	5.40	4.18	4.32	5.88	9.69	-	-	-	-
5	15.72	9.12	6.17	5.40	6.42	9.84	-	-	-	-	-
6	16.61	10.23	7.72	7.83	10.81	-	-	-	-	-	-
7	18.39	12.45	10.82	13.08	-	-	-	-	-	-	-
8	21.94	16.88	17.79	-	-	-	-	-	-	-	-
9	29.04	27.28	-	-	-	-	-	-	-	-	-
10	46.30	-	-	-	-	-	-	-	-	-	-

4.5 Architecture Analysis for the 10-Bit Folding and Interpolating A/D Converter

The input optimization of the system resolution carried out in Chapter 4.3.1 was based only on the total number of comparators since this gave us a first clue about the converter complexity. However, this kind of optimization is only a rough prediction since there are not just comparators in the folding and interpolating A/D converter. Fitting results from Chapters 4.4.2 and 4.4.3 into equations described in Chapters 4.2.2 and 4.2.3 can get a refined optimization. The combination of different single stages of the folding and interpolating A/D converter can be now optimized.

4.5.1 Die Area

According to Eqs. (3.35) and (3.40), the coarse resolution m, the folding resolution n_1, and the interpolating resolution n_2 are the main converter parameters that can be chosen by the designer. Aiming at the total resolution of the A/D converter to be 10-bit, the degree of freedom shrinks to two. The last parameter, the interpolating resolution n_2, can be determined by the first two parameters, m and n_1. Based on Eq. (4.10) and the scaling units in Table 4.2, the estimated die area of all combinations of a 10-bit folding and interpolating A/D converter can be obtained by varying the two parameters, m and n_1, from 0 to 10.

Table 4.11 describes the results of the parameter variations for the folding and interpolating A/D converter in respect to the die area. The case

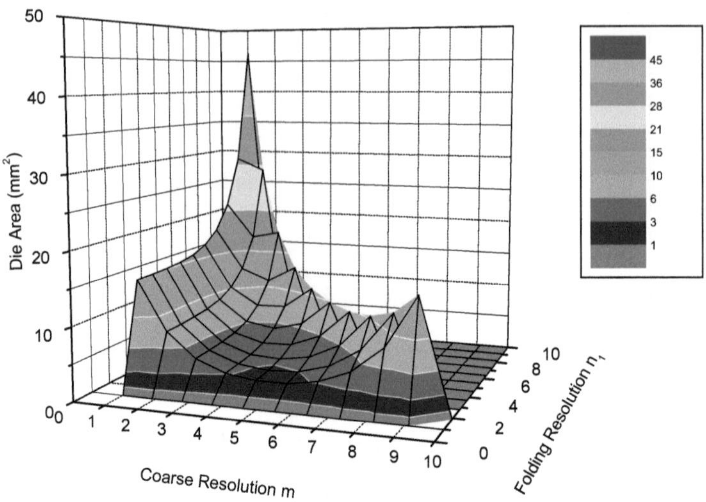

Fig. 4.14: Die area estimate for a 10-bit A/D converter

$(m = 10) \wedge (n_1 = 0)$ is a special case since the converter is no more a folding and interpolating A/D converter, but a flash A/D converter. Except for this, there is no other architecture that has folding resolution of 0. The smallest die area is obtained if the coarse resolution m equals to 4 and the folding and the interpolating resolutions, n_1 and n_2, equal to 1 and 5, respectively. The die area for this 4/1/5-architecture amounts to $2.49\ mm^2$.

The die area of 4/3/3-architecture which was found to be the best in Chapter 4.3.1 is indicated with an "*" in the table. The various architectures of the same interpolating resolution n_2 are positioned on the diagonal from bottom-left to top-right.

Fig. 4.14 illustrates the distribution of the die area estimated for various architectures of the 10-bit folding and interpolating A/D converter as a three dimensional plot. This plot shows how the die area of the 10-bit folding and interpolating A/D converter can be minimized.

Table 4.12: Estimate of the power dissipation [mW] as a function of parameter variations

	Coarse Resolution: m										
Folding Resolution: n_1	0	1	2	3	4	5	6	7	8	9	10
0	-	-	-	-	-	-	-	-	-	-	307.3
1	240.6	130.6	76.2	50.3	39.9	39.7	49.8	75.0	128.1	235.8	-
2	241.7	132.0	78.2	53.4	45.3	49.6	68.6	111.8	201.1	-	-
3	244.1	134.9	82.2	59.7	56 *	69.3	106.3	186.0	-	-	-
4	248.7	140.7	90.3	72.2	77.5	108.8	182.8	-	-	-	-
5	258.1	152.3	106.4	97.3	120.5	190.0	-	-	-	-	-
6	276.8	175.5	138.5	147.4	211.3	-	-	-	-	-	-
7	314.2	221.9	202.8	257.2	-	-	-	-	-	-	-
8	389.0	314.6	350.7	-	-	-	-	-	-	-	-
9	538.6	538.5	-	-	-	-	-	-	-	-	-
10	914.6	-	-	-	-	-	-	-	-	-	-

4.5.2 Power Dissipation

Concerning the power dissipation, a similar method has been used to estimate the power dissipation of various architectures of the 10-bit folding and interpolating A/D converter as in previous section although only the power of the analog part has been taken into consideration. The results are summarized in Table 4.12.

The power dissipation of 4/3/3-architecture from Chapter 4.3.1 is indicated with an "*" in the table and has an amount of 56 mW. The extreme example of $(m = 10) \wedge (n_1 = 0)$ describes the estimated power dissipation of a 10-bit flash A/D converter.

It can be found from this table that there are some architectures whose power dissipation is less than the 4/3/3-architecture. For example, the power dissipation is minimized for coarse resolution of $m = 5$ and the folding resolution of $n_1 = 1$, i.e. for 5/1/4-architecture. Its power dissipation is 39.7 mW, which is 16.3 mW and thus less than the 4/3/3-architecture. Only the folding factor F_F for this arrangement is 32. That is to say, the internal frequency inside the folding stage is very high, and the converter performance can be degraded. Another example is the 4/1/5-architecture, which consumes 39.9 mW, but this architecture has 5-bit as its interpolating resolution. This leads to larger interpolating error. Hence, Table 4.12 shows us that there is a refined optimization based on the power dissipation. Only careful designs to avoid high internal frequency or interpolating error are still necessary.

Fig. 4.15 illustrates the distribution of the power dissipation estimated for various architectures as a three-dimensional plot. This plot shows how the

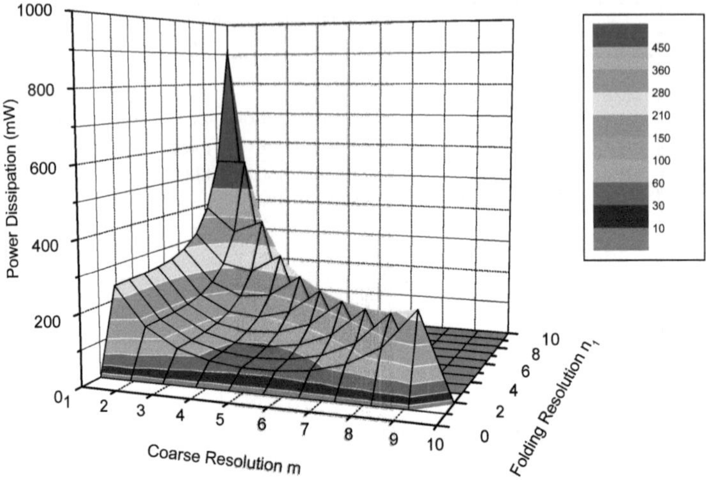

Fig. 4.15: Estimate of the power dissipation for a 10-bit A/D converter

power dissipation of the 10-bit folding and interpolating A/D converter can be minimized.

The discussion above suggests two principles for optimization of folding and interpolating A/D converters based either on the total die area or the power dissipation of the conversion system. The results can be summarized as follows:

- The architectural analysis concerning die area and power dissipation shows that architectures with larger die areas consume also more power.
- For a fixed the coarse resolution *m*, both the die area and the power dissipation increase while the folding resolution n_1 increases.
- For a fixed the folding resolution n_1 there is a minimum in respect to the die area and the power dissipation while the coarse resolution *m* is near the mid point of its range.

4.6 Architectural Hard Scaling

Resolution scaling of flash A/D converters is usually given by the total number of the representable states. This rough rule of thumb does not apply to A/D converters implemented in folding and interpolating architecture. The situation here is much more complex than the case of the flash A/D converter. The use of modules for the architecture of folding and interpolating A/D converters as shown above enables a quick estimate of the die area and the power dissipation for each particular resolution.

Apart from the architectural analysis of the 10-bit folding and interpolating A/D converter, the architectural hard scaling for other resolutions has also been under investigation in this book. The investigation of the architectures for the 8- and 12-bit A/D converters yields a high-speed converter family with a resolution between range 8~12 bit.

4.6.1 Architecture Investigation for an 8-Bit A/D Converter

Table 4.13 and Fig. 4.16 show the estimate for the die area of all possible combinations between the coarse resolution m and the folding resolution n_1 of an 8-bit A/D converter. Similarly to the 10-bit version, the combination $(m = 8) \wedge (n_1 = 0)$ indicates the special case of the flash converter and is not illustrated in the three dimensional plot.

The best five combinations with respect to the die area are shown and indicated in bold italics in the table. Except for the 4/2/2-architecture, the interpolating resolution n_2 of all other four combinations is higher than 2 bits. To avoid the interpolation error, lower n_2 is more favorable. Thus, the 4/2/2-architecture represents the best trade-off in respect to the die area.

Table 4.14 and Fig. 4.17 describe the performance in respect to the power dissipation estimated for all combinations of the 8-bit folding and interpolating A/D converter.

The best five combinations are distinguished from other combinations and indicated in bold italics in the table. Regarding the constraint that the interpolating resolution n_2 should not be too high, the 5/1/2-architecture would seem to be the best trade-off. Compared to the 4/2/2-architecture, the power dissipation increases only by $0.35\ \mathrm{mW}$, but the resolution distribution is better for the 4/2/2-architecture. The internal frequency can be reduced by factor 2. As a result, the 4/2/2-architecture is still the best trade-off in respect to the power dissipation.

Table 4.13: Estimate of the die area for an 8-bit A/D converter

mm^2	Coarse Resolution: m								
Folding Resolution: n_1	0	1	2	3	4	5	6	7	8
0	-	-	-	-	-	-	-	-	6.49
1	4.64	2.95	***2.15***	***1.84***	***1.84***	2.18	3.01	4.76	-
2	4.69	3.02	2.25	***1.99***	***2.11***	2.66	3.95	-	-
3	4.80	3.16	2.44	2.29	2.63	3.65	-	-	-
4	5.02	3.44	2.83	2.90	3.73	-	-	-	-
5	5.47	3.99	3.61	4.21	-	-	-	-	-
6	6.36	5.10	5.35	-	-	-	-	-	-
7	8.13	7.70	-	-	-	-	-	-	-
8	12.45	-	-	-	-	-	-	-	-

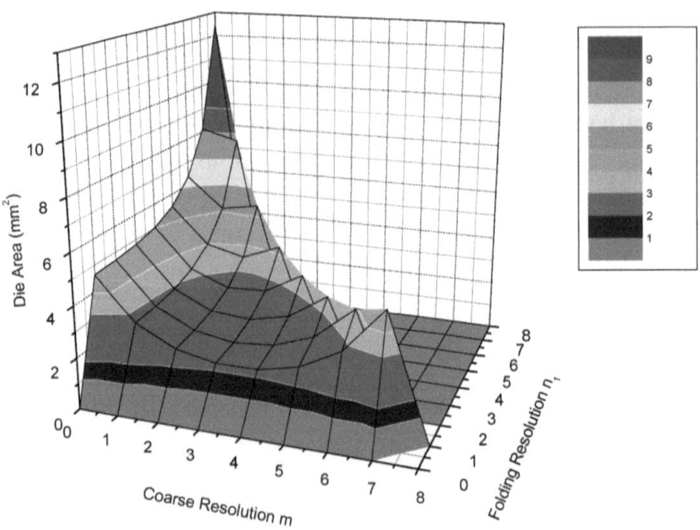

Fig. 4.16: Die area estimate for an 8-bit A/D converter

Table 4.14: Estimate of the power dissipation for an 8-bit A/D converter

mW		Coarse Resolution: m								
		0	1	2	3	4	5	6	7	8
Folding Resolution: n_1	0	-	-	-	-	-	-	-	-	91.21
	1	74.95	47.77	***34.82***	***29.60***	***29.53***	***34.55***	47.17	73.87	-
	2	76.12	49.22	36.83	***32.73***	34.90	44.41	66.31	-	-
	3	78.46	52.12	40.85	39.00	45.65	64.73	-	-	-
	4	83.13	57.92	48.89	51.52	68.35	-	-	-	-
	5	92.48	69.51	64.97	78.98	-	-	-	-	-
	6	111.18	92.70	101.93	-	-	-	-	-	-
	7	148.58	148.67	-	-	-	-	-	-	-
	8	242.58	-	-	-	-	-	-	-	-

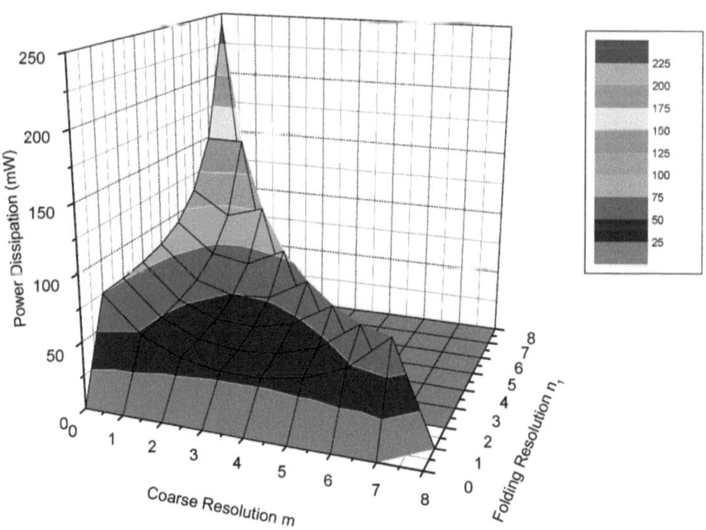

Fig. 4.17: Power dissipation estimate for an 8-bit A/D converter

Table 4.15: Estimate of the die area for a 12-bit A/D converter

mm^2	Coarse Resolution: m												
Folding Resolution: n_1	0	1	2	3	4	5	6	7	8	9	10	11	12
0	-	-	-	-	-	-	-	-	-	-	-	-	86.06
1	46.73	25.53	14.98	8.25	***5.05***	***3.78***	***3.81***	5.15	8.47	15.44	29.53	57.79	-
2	49.85	28.67	15.07	8.40	5.31	***4.27***	***4.74***	6.96	12.05	22.54	43.71	-	-
3	56.09	28.80	15.27	8.70	5.84	5.23	6.59	10.58	19.20	36.78	-	-	-
4	56.31	29.08	15.65	9.31	6.89	7.16	10.28	17.81	33.54	-	-	-	-
5	56.75	29.64	16.43	10.53	8.99	11.03	17.68	32.37	-	-	-	-	-
6	57.64	30.74	17.98	12.96	13.19	18.76	32.67	-	-	-	-	-	-
7	59.42	32.96	21.08	17.83	21.58	34.61	-	-	-	-	-	-	-
8	62.97	37.39	27.28	27.56	39.15	-	-	-	-	-	-	-	-
9	70.07	46.26	39.68	48.55	-	-	-	-	-	-	-	-	-
10	84.27	63.99	67.54	-	-	-	-	-	-	-	-	-	-
11	112.67	105.58	-	-	-	-	-	-	-	-	-	-	-
12	181.72	-	-	-	-	-	-	-	-	-	-	-	-

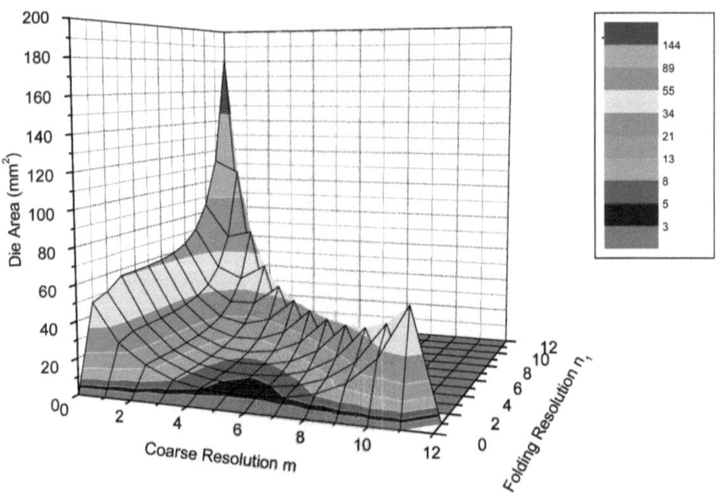

Fig. 4.18: Die area estimate for a 12-bit A/D converter

Table 4.16: Estimate of the power dissipation for a 12-bit A/D converter

mW	Coarse Resolution: m												
Folding Resolution: n_1	0	1	2	3	4	5	6	7	8	9	10	11	12
0	-	-	-	-	-	-	-	-	-	-	-	-	1171.7
1	672.6	385.0	241.8	133.1	81.3	***60.4***	***60.1***	***80.2***	130.7	236.9	451.9	883.3	-
2	750.5	463.2	243.8	136.2	86.7	***70.3***	***78.9***	117.0	203.4	381.4	740.2	-	-
3	906.5	466.1	247.8	142.5	97.4	90.0	116.6	190.5	348.7	670.9	-	-	-
4	911.1	471.9	255.9	155.0	118.9	129.5	192.0	337.7	640.6	-	-	-	-
5	920.5	483.5	272.0	180.1	161.9	208.3	342.6	634.3	-	-	-	-	-
6	939.2	506.7	304.1	230.2	247.9	366.1	648.8	-	-	-	-	-	-
7	976.6	553.1	368.4	330.4	419.9	691.3	-	-	-	-	-	-	-
8	1051.4	645.8	497.1	530.8	783.1	-	-	-	-	-	-	-	-
9	1201.0	831.3	754.4	970.1	-	-	-	-	-	-	-	-	-
10	1500.2	1202.3	1345.7	-	-	-	-	-	-	-	-	-	-
11	2098.6	2097.9	-	-	-	-	-	-	-	-	-	-	-
12	3602.6	-	-	-	-	-	-	-	-	-	-	-	-

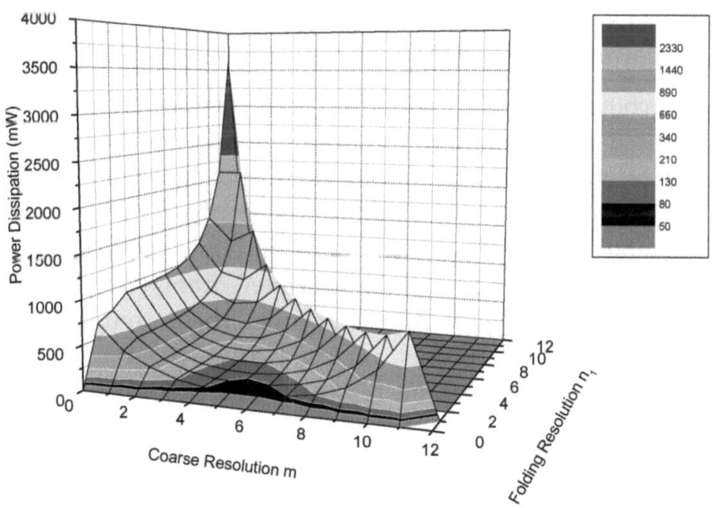

Fig. 4.19: Power dissipation estimate for a 12-bit A/D converter

4.6.2 Architecture Investigation for a 12-Bit A/D Converter

Table 4.15 and Fig. 4.18 show the estimates of the die area for a 12-bit folding and interpolating A/D converter that have been carried out in the same way as previous estimates. Both the coarse resolution m and the folding resolution n_1 vary from 0 to 12.

It can be observed from the plot that the best trade-off in respect to die area is the 6/2/4-architecture. Although the 6/3/3-architecture is more "traditional" one, its chip die area is 40% larger than the chip area of the 6/2/4-architecture.

Considering the power dissipation, the values are summarized both in Table 4.16 and in Fig. 4.19. The best five architectures are indicated again in bold italics. It can be seen that the 6/2/4-architecture is still the favorite architecture when power dissipation is considered. The straightforward 6/3/3-architecture dissipates 48% more power than the 6/2/4-architecture.

4.7 Summary

The work in this chapter has shown how the architecture of the folding and interpolating A/D converter can be optimized with respect to the die area and the power dissipation.

The architecture of the folding and interpolating A/D converter has been decomposed into a number of single components. General forms of the predictions of the die area and power dissipation can be seen in Eqs. (4.10) and (4.15). The coefficients are influenced by the distribution of the coarse, folding, and interpolating resolutions and the variables themselves are determined by empirical values measured from an implemented 10-bit folding and interpolating A/D converter. These components have been characterized and the results are used as basis for estimate of various combinations of coarse and fine resolutions.

This method has been also used to optimize architectures of 10-bit A/D converter realized in this book. A post-layout simulation has been carried out to estimate the effects of parasitic components and thus to ensure the best performance of the A/D converter [104].

A prototype of the 10-bit folding and interpolating A/D converter has been implemented in $0.6\ \mu m$ CMOS technology and the core area amounts $3.8125\ \text{mm}^2$. As discussed in Chapter 4.3, while the sampling rate is at

40 MHz, the analog and digital parts dissipate 20 mA and 12 mA at 3.3 V power supply voltage, respectively. If the sampling rate reduces to 20 MHz, the analog and digital parts dissipate 19 mA and 7 mA at 3.3 V, respectively. The applied technology is a standard CMOS technology, which can be characterized by the minimum channel length of 0.6 μm, three metal layers, and one low-ohmic polysilicon layer. A high-ohmic polysilicon layer was added for realization of the resistor string.

The architectural scalability of the folding and interpolating A/D converter is discussed using the scaling units for die area, power dissipation, and input capacitance of interstages. Prediction results of extending the resolution to 12-bit and shrinking to 8-bit are shown in Chapter 4.6.

CHAPTER 5

PHYSICAL DESIGN OF LOW-POWER, HIGH-SPEED EMBEDDED CMOS A/D CONVERTER

Epitome

5.1 Power Supply Network

Creating a "good" power supply network is a must for any circuits design and technology. That is, this includes any level, e.g. chip, package, or PCB.

Powering active devices requires DC voltage sources that are in practice never ideal. These nonidealities (e.g. finite source impedance) and interconnect properties (e.g. finite conductor wire or trace inductance) may cause unwanted communications between different devices that share the same power supply voltage. Hence, it is mandatory to take measures, to reduce these communications because they may degrade circuit performance due to noise from other circuits caused by power supply rail sharing. There are basically two techniques to achieve this: bypassing and decoupling.

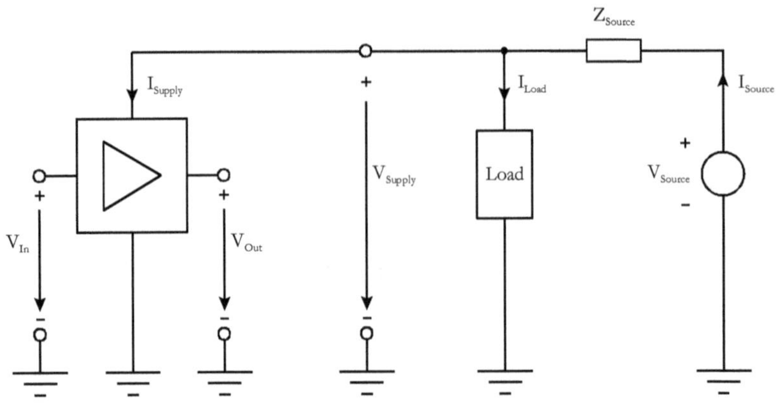

Fig. 5.1: Sharing a power supply rail

Bypassing techniques are based on trying to create a zero impedance at the power supply rails thus attempting to imitate an ideal voltage source. Decoupling techniques, on the other hand, rely on isolating adjacent noise sources at the power supply rails. Although both types of techniques often overlap in practice we shall investigate bypassing techniques first and add decoupling later. Also, we shall consider proper circuit design techniques that enable rejection of power supply noise.

The bypassing is necessary because any DC voltage source used to power a circuit necessarily exhibits non-zero source resistance and any interconnect that connects a circuit to this source exhibits a finite resistance and a finite inductance. The latter causes the interconnect impedance to rise at high frequencies and thus increases the danger of coupling interference into a circuit under consideration from other circuits sharing the same power supply rail. How far this interference causes problems, depends on the frequency response of the power supply rejection ratio (PSRR) of the circuit. To illustrate this consider the block diagram shown in Fig. 5.1 (although this applies to single power supply operation, it can be readily extended to double power supply operation).

In this figure we find an amplifier powered by a common supply voltage V_{Supply} and exhibiting a frequency dependent voltage gain of:

$$A_V(s) = \frac{V_{Out}(s)\big|_{V_{Supply}=const.}}{V_{In}(s)} \,. \tag{5.1}$$

Let us consider that the amplifier has a frequency-dependent PSRR defined as:

$$PSRR(s) = \frac{\dfrac{V_{Out}(s)\big|_{V_{Supply}=const.}}{V_{In}(s)}}{\dfrac{V_{Out}(s)\big|_{V_{In}=const.}}{V_{Supply}(s)}} . \tag{5.2}$$

Then the composite output voltage can be expressed as:

$$V_{Out}(s) = A_v(s) \cdot V_{In}(s) + \frac{A_v(s)}{PSRR(s)} \cdot V_{Supply}(s) , \tag{5.3}$$

owing to the principle of superposition. If we further assume that

$$A_V(s) = \frac{A_{V0}}{1+\dfrac{s}{p}} , \tag{5.4}$$

$$PSRR(s) = \frac{PSRR_{0Hz}}{1+\dfrac{s}{p_{PSRR}}} , \tag{5.5}$$

and

$$I_{Supply} \ll I_{Load} , \tag{5.6}$$

then we can write

$$V_{Out}(s) = A_V(s) \cdot \left[V_{In}(s) + \frac{\left[\dfrac{V_{Source} - I_{Load}(s) \cdot}{Z_{Source}(s)}\right] \cdot \left[1 + \dfrac{s}{p_{PSRR}}\right]}{PSRR_{0Hz}} \right] , \tag{5.7}$$

where V_{Source} is an ideal voltage source, $Z_{Source}(s)$ represents a combined source and interconnect impedance, and I_{Load} stands for current load caused by other circuits sharing the same supply voltage V_{Supply}. As mentioned above, the common loading of V_{Supply} can cause an undesired interference due to varying current I_{Load} (sometimes referred as noise at the power sup-

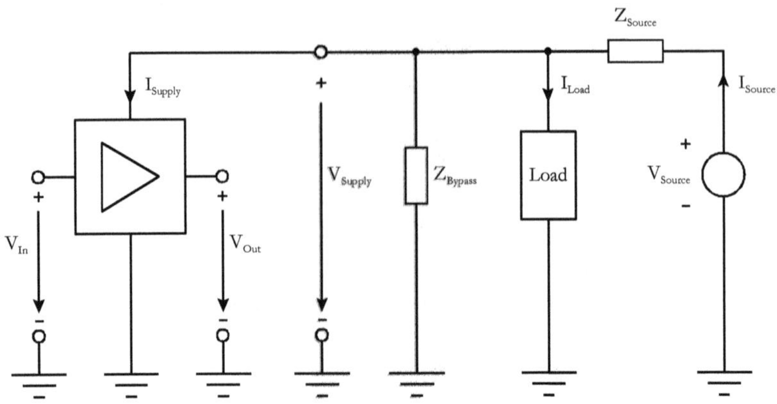

Fig. 5.2: Bypassing

ply rail) and degrade the amplifier performance at the output. This can be seen if we introduce

$$Z_{Source} = R_{Source} + s \cdot L_{Wire} \,, \tag{5.8}$$

where R_{Source} is the finite source resistance and L_{Wire} is the interconnect inductance (note that R_{Source} may also include wiring resistance R_{Wire}, if any). Then we obtain:

$$V_{Out}(s) = \frac{A_{VO} \cdot \left(1 + \dfrac{s}{p_{PSRR}}\right)}{1 + \dfrac{s}{p}} \cdot \left\{ V_{In}(s) + \frac{\left[\begin{array}{l} V_{Source} - I_{Load}(s) \cdot \\ R_{Source} \cdot \left(1 + \dfrac{s \cdot L_{Wire}}{R_{Source}}\right) \end{array} \right]}{PSRR_{0Hz}} \right\}, \tag{5.9}$$

which clearly suggest that at high frequencies the variations of the current $I_{Load}(s)$ affect greatly the output voltage $V_{Out}(s)$ due to zeroes located at p_{PSRR} and R_{Source}/L_{Wire}. Needless to say, that we should strive not only for high $PSRR_{0Hz}$, but also for high p_{PSRR} and low L_{Wire} (we cannot go for high R_{Source} as this would cause varying V_{Supply} at low frequencies for slowly varying I_{Load}).

A common remedy is the placement of a bypass capacitor between power and ground. In Fig. 5.2, this is represented by the general impedance Z_{Bypass}. For an ideal bypass capacitor C_{Bypass} we obtain:

$$V_{Out}(s) = \frac{A_{V0}}{1+\frac{s}{p}} \cdot \left\{ V_{In}(s) + \frac{\left[V_{Source} - I_{Load}(s) \cdot R_{Source} \cdot \left(1+\frac{s \cdot L_{Wire}}{R_{Source}}\right) \cdot \left(1+\frac{s}{p_{PSRR}}\right) \right]}{PSRR_{0Hz} \left(1 + s \cdot C_{Bypass} \cdot R_{Source} + s^2 \cdot L_{Wire} \cdot C_{Bypass} \right)} \right\}, \quad (5.10)$$

This expression shows that the bypass capacitor C_{Bypass} has introduced two more poles. Thus the behaviour at high frequencies improves as the noise effect due to varying $I_{Load}(s)$ is reduced. Nevertheless, C_{Bypass} forms together with L_{Wire} a parallel resonant circuit and this may adversely affect the amplifier operation at the resonance frequency if the amplifier is not capable to reject the supply noise at this frequency.

Thus while the bypass capacitor acts as an AC shunt and thus reduces the impedance between the power supply rails and the ground at high frequencies it may create problems at the resonance frequency.

The question is now where the resonance frequency should be located. To place it properly we should know the spectrum of the noise occurring at power supply rail but this is not easy to predict. Definitely, we should try to avoid that the resonance frequency $1/\sqrt{L_{Wire} \cdot C_{Bypass}}$ lies at switching frequencies occurring in circuits sharing the same power supply rail or their multiples. Nevertheless, even if we avoid this we must consider the frequency dependence of the $PSRR$.

An analysis of the above equation for the output voltage $V_{Out}(s)$ indicates that for a "good" PSRR it is required that

$$p \ll p_{PSRR} \ll \frac{R_{Source}}{L_{Wire}} . \quad (5.11)$$

The above mentioned resonance frequency should then lie below p_{PSRR}, i.e.

$$\frac{1}{\sqrt{L_{Wire} \cdot C_{Bypass}}} < p_{PSRR} \,, \tag{5.12}$$

and the bypass capacitor should fulfil the condition

$$C_{Bypass} > \frac{I_{Load,\,rms} \cdot SNR}{PSRR_{0Hz} \cdot p_{PSRR} \cdot V_{In,\,rms}} \,, \tag{5.13}$$

for a required signal-to-noise ratio (*SNR*).

Unfortunately, there are no ideal capacitors. To consider this we model a real bypass capacitor using

$$Z_{Bypass} = R_{Bypass} + s \cdot L_{Bypass} + \frac{1}{s \cdot C_{Bypass}} \,, \tag{5.14}$$

where R_{Bypass} and L_{Bypass} are finite series resistance and inductance, respectively. This yields an output voltage

$$V_{Out}(s) = \frac{A_{V0}}{1+\frac{s}{p}} \cdot \left\{ V_{In}(s) + \frac{\left\{ \begin{array}{l} \left[\begin{array}{l} 1 + s \cdot C_{Bypass} \cdot R_{Bypass} + \\ s \cdot L_{Bypass} \cdot C_{Bypass} \end{array} \right] \cdot \\ \left[\begin{array}{l} V_{Source} - I_{Load}(s) \cdot \\ R_{Source} \cdot \left(1 + s \cdot L_{Wire} \cdot R_{Source}^{-1} \right) \end{array} \right] \cdot \\ \left[1 + s \cdot p_{PSRR}^{-1} \right] \end{array} \right\}}{PSRR_{0Hz} \cdot \left[\begin{array}{l} 1 + s \cdot C_{Bypass} \cdot \\ \left(R_{Bypass} + R_{Source} \right) + \\ s \cdot \left(L_{Bypass} + L_{Wire} \right) \cdot \\ C_{Bypass} \end{array} \right]} \right\} , \tag{5.15}$$

which now contains two more zeroes while the number of poles has not increased. This suggests that we get into problems at high frequencies, as the bypass capacitor becomes less effective.

The conclusion at this point is clear: although a bypass capacitor is needed to lower the high frequency noise at power supply rails caused by other circuits, it can generate problems by creating a resonant frequency. Due to

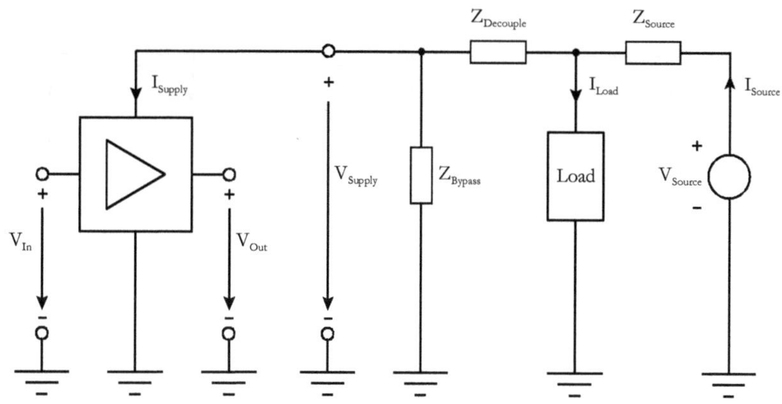

Fig. 5.3: Single bypassing and decoupling

this we have to codesign the power supply rejection ratio of the circuit powered and the bypass capacitor. Furthermore, the bypass capacitor should exhibit a very low series resistance and inductance to be effective at very high frequencies.

Similar conclusions apply to decoupling techniques. There are basically two techniques: active and passive decoupling. Active decoupling involves the use of voltage regulators, while passive decoupling is based on inserting either an inductor or a resistor into the power supply rail to filter all noise moving from any load to the circuit under consideration (see Fig. 5.3).

Let us now assume we employ a single bypassing. Then for

$$Z_{Decouple} = R_{Decouple} + s \cdot L_{Decouple} \ , \tag{5.16}$$

$$L_{Bypass} = \frac{1}{s \cdot C_{Bypass}} \ , \tag{5.17}$$

and

$$Z_{Source} = R_{Source} + s \cdot L_{Wire} \ , \tag{5.18}$$

we obtain

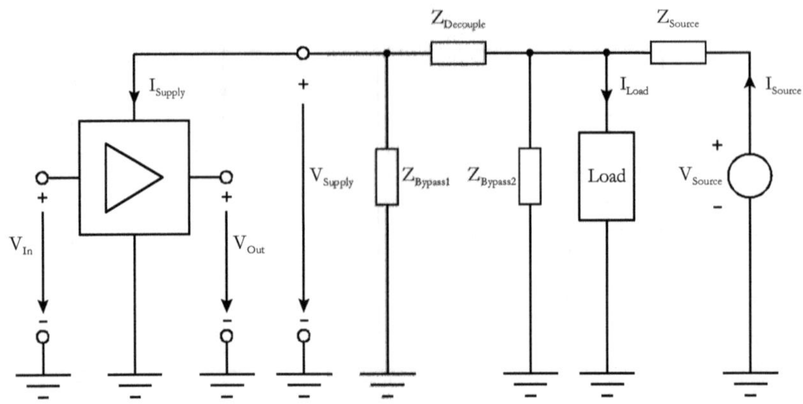

Fig. 5.4: Double bypassing and decoupling

$$V_{Out}(s) = \frac{A_{V0}}{1+\frac{s}{p}} \cdot \left\{ V_{In}(s) + \frac{\left[V_{Source} - I_{Load}(s) \cdot R_{Source} \cdot \left(1 + \frac{s \cdot L_{Wire}}{R_{Source}} \right) \right] \cdot \left(1 + \frac{s}{p_{PSRR}} \right)}{\left\{ PSRR_{0Hz} \cdot \left[1 + s \cdot C_{Bypass} \cdot \left(R_{Decouple} + R_{Source} \right) \right] + s \cdot \left(L_{Decouple} + L_{Wiere} \right) \cdot C_{Bypass} \right\}} \right\}, \tag{5.19}$$

again for negligible I_{Supply}. This suggests that single bypassing combined with decoupling does not yield any substantial improvement when compared with bypassing without decoupling since we again obtain only two more poles. Only double capacitive bypassing and inductive decoupling forming a π-network yields four more poles (Fig. 5.4), i.e. for

$$Z_{Bypass1} = \frac{1}{s \cdot C_{Bypass1}}, \tag{5.20}$$

$$Z_{Bypass2} = \frac{1}{s \cdot C_{Bypass2}}, \tag{5.21}$$

and

$$Z_{Decouple} = R_{Decouple} + s \cdot L_{Decouple}, \tag{5.22}$$

The use of a decoupling inductor, such as ferite toroid, may be a good idea for a discrete circuit, but in an integrated circuit it is impossible to implement high Q inductors. However, a small resistor $R_{Decouple}$ still helps to decouple. The output voltage is then

$$V_{Out}(s) = \frac{A_{V0}}{1+\frac{s}{p}} \cdot \left\{ V_{In}(s) + \frac{\left[V_{Source} - I_{Load}(s) \cdot R_{Source} \cdot \left(1 + \frac{s \cdot L_{Wire}}{R_{Source}} \right) \right] \cdot \left(1 + \frac{s}{p_{PSRR}} \right)}{PSRR_{0Hz} \cdot \left[1 + s \cdot \left(\begin{array}{l} \left(R_{Decouple} + R_{Source} \right) \cdot \\ C_{Bypass1} + C_{Bypass2} R_{Source} \end{array} \right) + s^2 \cdot \left(\begin{array}{l} L_{Wire} \cdot \left(C_{Bypass1} + C_{Bypass2} \right) + \\ C_{Bypass1} C_{Bypass2} R_{Source} R_{Decouple} \end{array} \right) + s^3 \cdot L_{Wire} C_{Bypass1} C_{Bypass2} R_{Source} \right]} \right\}, \quad (5.23)$$

again for the case when I_{Supple} is negligible. For high supply currents I_{Supple} and large $R_{Decouple}$, however, voltage drop across $R_{Decouple}$ would excessively decrease V_{Supply}.

To summarize our considerations, bypassing and decoupling represent important circuit design techniques to ensure "good" power supply, i.e. low power supply impedance at high frequencies. Nevertheless, these techniques also tend to introduce parallel resonance frequencies at which the power supply impedance sharply increases. If such a resonance frequency occurs at a "wrong" frequency, e.g. at frequency where *PSRR* of the powered circuit is reduced, these techniques can actually make things worse. Hence, they have to be applied with caution and cleverness.

5.2 Reference Voltage Network

Voltage reference circuits are often used in electronic systems to establish a system voltage reference. Unlike the case of voltage sources the main emphasis in a voltage reference circuit is not on the low output impedance but on the voltage precision and thermal stability. Nevertheless, there are other important design parameters, such as noise, line rejection, and load sensitivity. Especially voltage reference noise is often disregarded by designers although it plays an eminent role in design of A/D converters. An A/D converter converts an analog input voltage into a dimensionless digital code directly by comparing the input voltage with a reference voltage. Hence, for an N bit converter the total comparator and reference rms noise must be better than 0.5 LSB. This is quite difficult to achieve for high speed, high resolution A/D converters. On the one hand, since a voltage reference circuit has to provide only a DC precision voltage, its bandwidth - and thus its noise bandwidth, too - can be quite low for any converter. However, the voltage reference circuit must be also capable of rejecting noise from the power supply rails. Hence, voltage reference circuits used in high speed conversion must exhibit high PSRR (called often "line sensitivity" for voltage references and regulators), especially at high frequencies. This means that in practice we have to resort to bypassing and decoupling measures introduced in the previous section because we are facing similar problems.

As mentioned above, another important parameter is load sensitivity. Since the load sensitivity of most basic voltage reference stages is quite low, most of them have to be buffered using an output amplifier. Especially voltage reference circuits used in high speed converters suffer from heavy transient loading and thus require broadband buffer amplifiers.

Remaining voltage reference circuit parameters depend on its technological realization. First of all, in integrated A/D converters the key issue is whether to use an on-chip or an off-chip voltage reference stage. On-chip voltage reference generation can be based on forward-biased diode, zener diode, or bandgap principle or be derived from the power supply voltage using voltage dividers, e.g. resistor strings.

Simple forward-biased pn-diodes (or diode-connected bipolar transistors) present in bipolar or CMOS technologies exhibit high temperature coefficient (typ. -3000 ppm/K). In bipolar (or BiCMOS) their voltage is available in 0.6V jumps, while in CMOS it is only 0.6V (as diodes cannot be stacked). Zener diodes (using either zener or avalanche breakdown) yield breakdown voltages that are too high for circuits operating below 5V power supply voltage which are typical for today submicron technologies. Bandgap reference technique is the most common technique used in bipolar, BiCMOS, and CMOS technologies, as it operates at low voltages but the temperature coefficient (TC) can vary between 10 and 100 ppm/K, subject to design and technology. We can determine the required TC as:

$$TC\ [ppm/K] = \frac{10^6}{2^{N+1} \cdot \Delta T}, \tag{5.24}$$

where N is the converter resolution (in bits) and ΔT is the operating temperature range (in °C). Hence, TC=10ppm/K would allow only 9bit resolution over an operating temperature range of 100°C.

The use of attenuated power supply voltage (e.g. resistive voltage dividers) to generate a common reference voltage is a good idea, if a precisely defined power supply voltage is available but it requires extensive bypassing and decoupling since the line sensitivity of this technique is high. The resistor string used for voltage division must exhibit very high resistance to keep the quiescent current low (for low power consumption) but high resistance causes high white noise and requires buffering to eliminates loading effects.

Although many of chip voltage reference circuits (often available as commercial ICs) use the same principles as on-chip circuits (e.g. bandgap technique), there are some ICs on the market based on special technologies that may yield an improved performance. Thus the buried zener approach offers low TC (1-2ppm/K) and low noise. Special JFET devices, such as XFET, offer good performance as well (TC<10ppm/K). Many dedicated voltage reference ICs allow optional trimming and thus offer a better performance than on-chip voltage reference circuits anyway.

The use of external voltage reference ICs for A/D converter, however, harbours its perils, too. The chip-to-chip wiring and bond and lead frame wires increase the inductances in the power supply rail. Though this may help as in case of bypassing and decoupling, we obtain more parallel resonance frequencies. This again we have to make sure that these resonance frequencies do not cause any problems. Unfortunately, this is often the overlooked in practice.

Let us now briefly consider the requirements for voltage reference buffering. For this purpose consider the case shown in Fig. 5.5. It is similar to Fig. 5.1, but the voltage source V_{Source} has been replaced by a buffered voltage reference V_{Ref}.

If we assume that $A_{Buffer,\,0} \gg 1$, then we can readily derive that

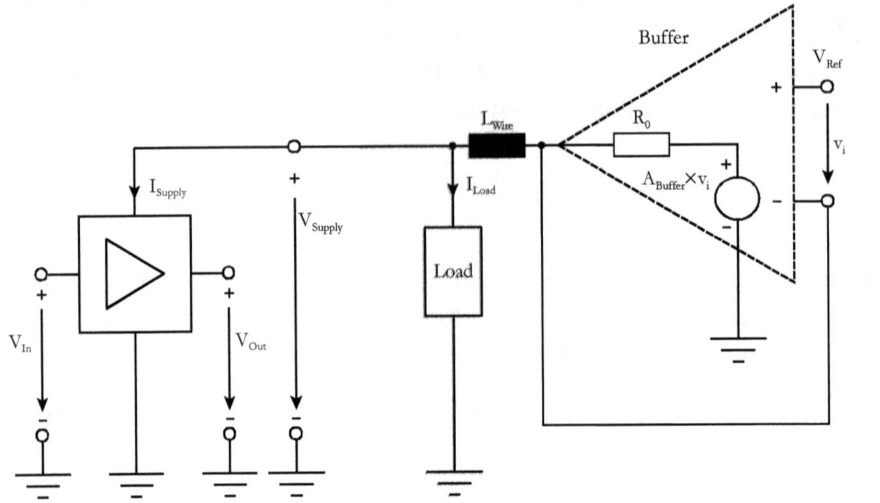

Fig. 5.5: Voltage reference buffering

$$V_{Source} \cong \frac{V_{Ref}}{1+\dfrac{s}{A_{Buffer,\,0} \cdot p_{Buffer}}}, \tag{5.25}$$

and

$$R_{Source} \cong \frac{R_0}{A_{Buffer,\,0}} \cdot \frac{1+\dfrac{s}{p_{Buffer}}}{1+\dfrac{s}{A_{Buffer,\,0} \cdot p_{Buffer}}}. \tag{5.26}$$

This is implying, that in the frequency range between p_{Buffer} and $A_{Buffer,\,0} \cdot p_{Buffer}$ the output impedance of the buffer exhibits an inductive behavior. This affects the output voltage $V_{Out}(s)$, which can now be given as:

$$V_{Out}(s)=\frac{A_{V0}}{\left(1+\frac{s}{p}\right)}\cdot\left\{V_{In}(s)+\frac{\left\{\begin{array}{l}V_{Ref}-\frac{I_{Load}(s)\cdot R_0}{A_{Buffer,\,0}}\cdot\left(1+\frac{s}{p_{PSRR}}\right)\\ \left[\begin{array}{l}1+\frac{s^2\cdot L_{Wire}}{p_{Buffer}\cdot R_0}+\\ s\cdot\left(\frac{1}{p_{Buffer}}+\frac{A_{Buffer,\,0}\cdot L_{Wire}}{R_0}\right)\end{array}\right]\end{array}\right\}}{PSRR_{0Hz}\cdot\left(1+\frac{s}{p_{Buffer}\cdot A_{Buffer,\,0}}\right)}\right\}. \tag{5.27}$$

This suggest that the use of narrowband buffer increases the inductance at the power supply rail already at frequencies above p_{Buffer} and thus the use of a bypass capacitor is a must (see Fig. 5.6).

Its size in this case, however, must have a suitable value, since the bypass capacitor still tends to resonate with the wiring inductance, although this may be quite low. This effect can be seen from the expression for the output voltage

$$V_{Out}(s)=\frac{A_{V0}}{1+\frac{s}{p}}\cdot\left\{V_{In}(s)+\frac{\left[V_{Ref}-\frac{I_{Load}(s)\cdot R_0}{A_{Buffer,\,0}}\left(\begin{array}{l}1+\frac{s}{p_{Buffer}}+\\ \frac{sA_{Buffer,\,0}L_{Wire}}{R_0}+\\ \frac{s^2L_{Wire}}{p_{Buffer}R_0}\end{array}\right)\right]}{PSRR_{0Hz}\left[\begin{array}{l}\frac{\frac{s}{p_{Buffer}}+sC_{Bypass}R_0}{A_{Buffer,\,0}}+\\ s^2L_{Wire}C_{Bypass}+\\ \frac{s^3L_{Wire}C_{Bypass}}{p_{Buffer}A_{Buffer,\,0}}+1\end{array}\right]\left[1+\frac{s}{p_{PSRR}}\right]^{-1}}\right\}. \tag{5.28}$$

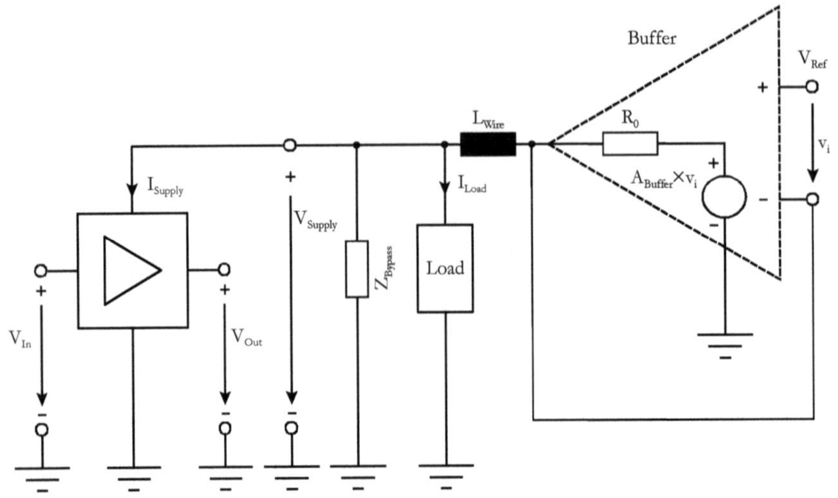

Fig. 5.6: Voltage reference buffering with a bypass

To keep our considerations simple we have not considered the PSRR of the buffer, but in real chip design this must be considered as well. The same applies to the decoupling which can be used here, too.

We can conclude by stating that the use of a buffered voltage reference requires a careful design of the voltage reference buffer and of the power supply rejection ratios of the powered circuits and as well as bypassing and decoupling measures. The best approach seems to be a proper placement of zeros and poles of the transfer function (under consideration of $PSRR_{0Hz}$) to ensure the maximum line noise rejection, especially at resonance frequencies.

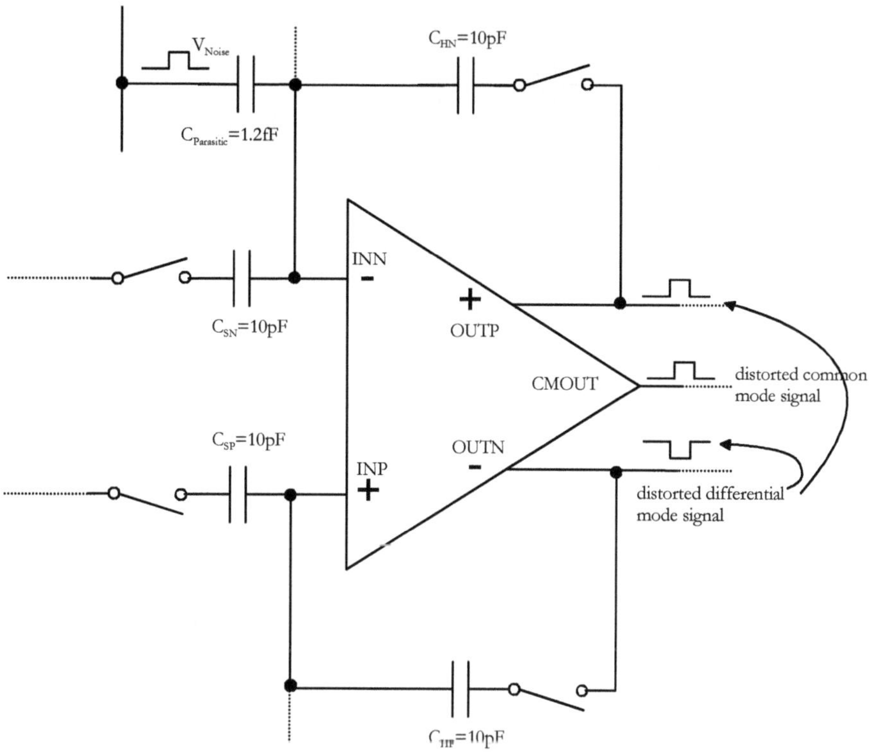

Fig. 5.7: Capacitive noise coupling on sample & hold input node

5.3 Noise Coupling

In this chapter we will discuss the effects of substrate crosstalk in ICs and describe some of the measures to combat it. The discussion contains results of our own deliberations and past experience as well as reflections on published literature on this topic.

In typical SoC with embedded ADCs, about 85% - 95% of the total chip area is digital circuitry, while 10% belongs to analogue functional blocks. The ADC itself contains about 60% digital cells, e.g. the error correction, the thermometer code converter, the clock generators, and the comparators. To save cost, area, and power consumption, a single chip solution is still the best standard system implementation although this high integration as a mixed-signal ASIC has technical disadvantages, which take additional effort to cope with.

Substrate noise/crosstalk and noise coupling are two of the key problems in mixed-signal analog/digital ICs. In the following chapters different kinds

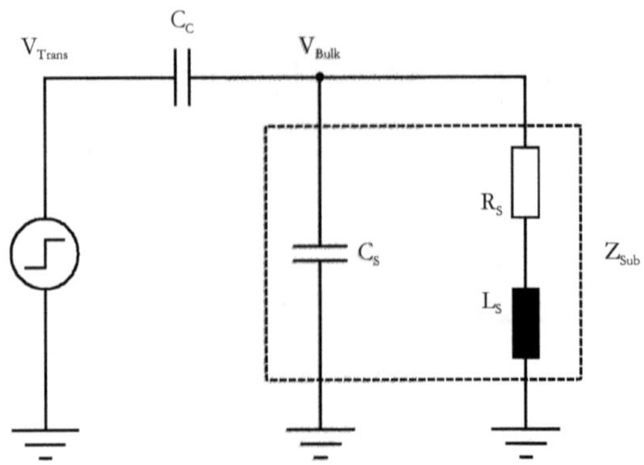

Fig. 5.8: Comprised model for digital switching noise in heavily doped substrate

of coupling and crosstalk mechanisms are investigated and appropriate design measures are defined to minimize ADCs performance degeneration.

5.3.1 Inductive Noise Coupling

Chip wires, bond wires, and package leads all exhibit self- and mutual-inductance which induce signal and power bounce (together with parasitic capacitances they form resonance circuits, see below) and also cause interference coupling due to transformer effect. Bouncing voltage spikes become quite large even at moderate currents and reach some hundreds of milivolts under worse conditions. The inductive coupling is easily exceeded by capacitive coupling for higher resistance circuit nodes. The inductive type of coupling is most effective at nodes exhibiting low resistance. For this reason, the inductance of bond wires and package leads diminishes an effective capacitive off-chip decoupling of the power supply lines.

Whereas power bounces can effect all circuit blocks and all signals at once, the transformer effect, which is based on the mutual inductance coupling, only affects loops of neighboring signals. The most obvious way of mutual inductive coupling is the inductive coupling of two neighboring bond wires. But also overlaying signal loops of on chip wiring can induce signal distortions, especially if digital and analog signal loops are not separated properly in all metal planes.

The most effective way of reducing mutual inductive coupling is a separation of sensitive signals from disturbing signal loops and a reduction of the loop area. Since spatial separation requires costly chip area, the most effective way of mutual inductance reduction is by antiparallel signal path routing. For on chip wiring this also is a practical way to reduce the risk of accidentally formed overlaying signal loops.

5.3.2 Capacitive Coupling

Unlike inductive coupling, capacitive coupling poses a severe problem at high resistance nodes. Hence, the capacitive coupling in the presence of a low resistance ground or chip substrate and at higher levels of metal is different. For lower metal levels coupling to and from the substrate is more important than on higher metal levels. But in today's deep submicron processes the parasitic sidewall capacitance usually exhibits higher values than coresponding vertical capacitance of the same size. Furthermore, the value of sidewall capacitance is even increasing for higher metal levels. Therefore, a good choice of the metal level is the first step to prevent noise injection to the substrate and to choose carefully which nodes are neighboring is the first step to prevent crosstalk.

For todays high resolution analogue signal processing it is essential to recognize all possible ways of parasitic capacitive coupling into high impedance nodes, even though the parasitc capacitances may be only in the picofarad range.

Consider the differential-path sample-and-hold shown in Fig. 5.7 with 10 pF hold capacitors as an example. For a sample-and-hold circuit with N bit resolution the difference of parasitic capacitances at the high impedance input nodes (INP and INN) should be below:

$$\Delta C_{Parasitic} \leq 2 \cdot \frac{V_{Out} \cdot C_{Hold}}{V_{Noise} \cdot 2^{N}} . \tag{5.29}$$

For a 14 bit resolution signal with 2 V output swing and a digital noise source of 2.5 V, all the parasitic capacitance should be below 1.2 fF. Thus even though differential-path signal processing should be more or less immune against common-mode crosstalk, it is clear to see from the values above, that any path mismatch can cause non-linearity.

Hence, the use of parasitic extraction tools and the simulation of the parasitic extracted netlist is strongly recommended for high performance analogue signal processing, as it is used in nowadays embedded analogue-to-digital converters.

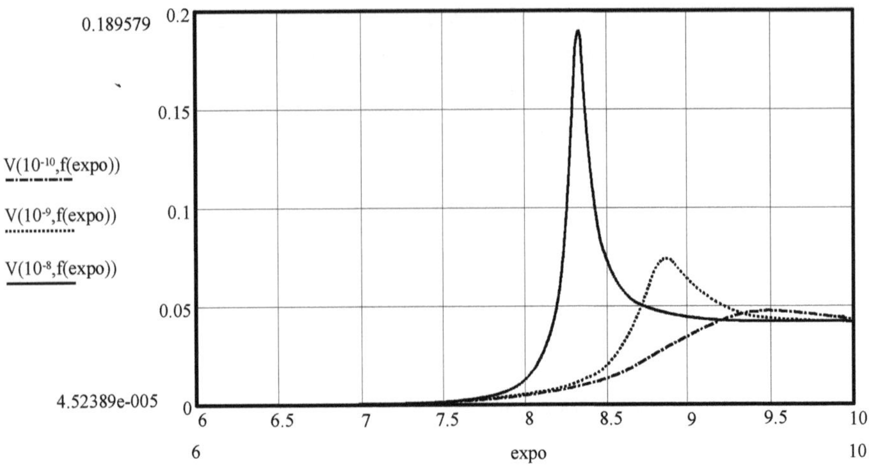

Fig. 5.9: Magnitude of transfer function for different inductances

5.3.3 Substrate and Package Coupling

The power supply rails of an IC generally have the most capacitance to chip substrate and to each other due to the large number of devices connected to them. This large capacitance together with the inductance and resistance from the chip substrate, bond wires, and package leads forms multiple RLC resonance circuits with their own distinct natural frequencies. Any inductive and capacitive coupling, however, will affect these resonance frequencies and must be accounted for. The resonance behavior depends of the Q of the system. Tuning of the resonance frequencies is possible by controlling the amount of on-chip capacitance and off-chip inductance. Balancing the capacitances between power and ground rails to substrate is one possibility to cancel the capacitive current injected to the substrate to the first order. Normally the digital power supply to substrate capacitance is the largest in standard SoC due to the large n-well to substrate area used for the pmos transistors in the digital circuitry.

When modeling and simulating substrate crosstalk we can use the model in Fig. 5.8. In [108], the results were found to be highly dependent on Cs (capacitance between bulk and IC package cavity), Rs (spreading resistance from surface substrate contacts to bulk node) and Ls (inductance, determined by the number of bonding pads and package pins used to bias the substrate contacts at the die surface). The simplified schematic (see Fig. 5.8) including Cc (models diffusion and interconnect capacitances coupling the switching noise sources, e.g. noisy power supply lines, to the substrate) and a switching transient source V_{trans} yields the following transfer function for the substrate voltage V_{bulk}:

$$\frac{V_{Bulk}}{V_{Trans}} = \frac{C_C}{C_C + C_S} \cdot \frac{s \cdot (s + \frac{R_S}{L_S})}{s^2 + \frac{R_S}{L_S} s + \frac{1}{L_S (C_C + C_S)}} \,. \quad (5.30)$$

The resonance frequency is defined as:

$$\omega_0 = \sqrt{\frac{1}{L_S (C_S + C_C)} - \frac{R_S^{\,2}}{L_S^{\,2}}} \,. \quad (5.31)$$

Fig. 5.9 shows the magnitude of the above transfer function for three different inductance values: 0.1nH (red bottom curve), 1nH (blue middle curve) and 10nH (green top curve). The x-axis represents an exponentially scaled frequency going from 1MHz to 10GHz.

For a given Rs, Cs, and Cc the magnitude of the frequency response is very high for a large Ls (see Fig.2). Care must be taken to ensure that all switching frequencies and their low-order harmonics do not coincide with the substrate resonance frequency. Calculations show that the amplitude of substrate noise can be reduced by decreasing the value of Cc with respect to Cs and by decreasing the value of Ls with respect to Rs. Unfortunately, these values cannot be changed arbitrarily, e.g. because of dependence on technology parameters. Increasing Cs by adding on chip decoupling capacitances lowers the resonance frequency. The best practical approach is to reduce Ls with respect to Rs. This will be discussed in the following two chapters.

5.3.3.1 Reducing Interconnect Inductance

As it was shown in the previous chapter, the most effective way for a reduction of substrate and package introduced noise is, to reduce the interconnect inductance.

The consideration of package introduced noise coupling should already start in the early design phase, by choosing the most suitable IC package. As it is to see in Table 5.1, in terms of interconnect inductances, standard IC packages can vary over more than 3 decades. For a given number of pins, it is recommended to choose always the package with lowest values of interconnect inductances, if high performance analogue functionality should be integrated in a SoC. It is also to see, that the interconnect inductance can be reduced by choosing the right pin location in a given package. In Table 5.1 it is shown that the center pins of standard IC packages, exhibit a much lower inductance than the corner pins. Hence these pins should be reserved for the high precision analogue functional block IOs.

A further reduction of interconnect inductances can be achieved by using multiple IO schemes. Very often it is observed, that multiple parallel bondwires are used to reduce the bondwire inductance. Due to the influence of the mutual inductance, which has to be added to the self inductance for parallel wires, this scheme is not the most effective one in terms of reduction of the interconnect inductance. As it is to see in Fig. 5.10, that the mutual inductance can be subtracted from the self inductance if an antiparallel IO scheme is used.

Under realistic IC package conditions, the inductances for a 3 mm long bondwire can be calculated as follows:

Table 5.1: Pin inductances of standard IC packages

Package	Max. self inductance	Min. self inductance	Max. mutual inductance	Min. mutual inductance
DIL 28 (w/o ground plane [109])	50 nH (corner)	3 nH (center)	-	-
SOIC 28 [109]	3.76 nH (corner)	2.32 nH (center)	1.38 nH (corner)	0.98 nH (center)
TQFP 64 pin [109]	5.56 nH (corner)	3.21 nH (center)	3.89 nH (corner)	2.02 nH (center)
PGBA [109]	10.9 nH (corner)	6.5 nH (center)	1 nH (center)	0.4 nH (corner)
Flip chip (standard) [110]	2.16 nH	0.26 nH	-	-
Flip chip (micro lead frame) [111]	> 0.51 nH	> 0.05 nH	-	-

$$\frac{L_{Self}}{l} = \frac{\mu_r \cdot \mu_0}{2\pi} \cdot \ln\left(\frac{2h}{r}\right) \left[\frac{\text{Henries}}{\text{Meter}}\right] \qquad \frac{L_{Mutual}}{l} = \frac{\mu}{4\pi} \cdot \ln\left[1 + \left(\frac{2h}{d}\right)^2\right] \left[\frac{\text{Henries}}{\text{Meter}}\right]$$

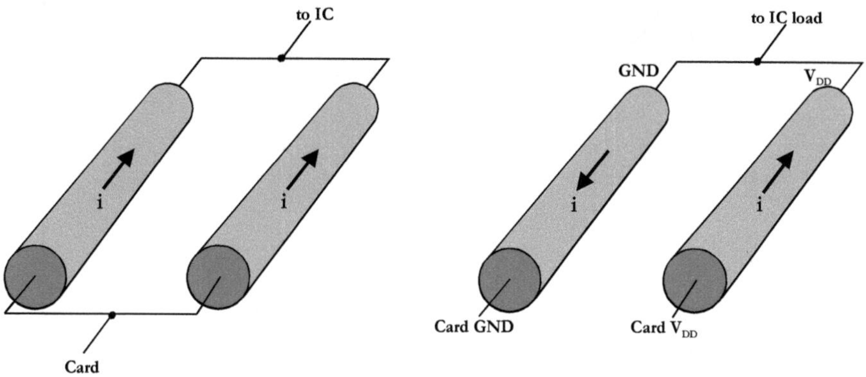

$$L_{Eff} = \frac{\left(L_{Self} + L_{Mutual}\right)}{2} \qquad L_{Eff} = 2 \cdot L_{Self} - 2 \cdot L_{Mutual}$$

Fig. 5.10: Inductance of parallel and antiparallel bondwires.

$$\frac{L_{Self}}{l} \cong 0.2 \cdot \ln\left(\frac{2h}{r}\right) \cong 0.2 \cdot \ln\left(\frac{2 \cdot 2mm}{0.015mm}\right) \cong 1.11 \frac{nH}{mm}, \tag{5.32}$$

for 3mm bond wire $L_{Self} \cong 3.33\ nH$ and

$$\begin{aligned} \frac{L_{Mutual}}{l} &\cong 0.1 \cdot \ln\left[1 + \left(\frac{2h}{d}\right)^2\right] \\ &\cong 0.1 \cdot \ln\left[1 + \left(\frac{2 \cdot 2mm}{0.1mm}\right)^2\right] \simeq 0.74 \frac{nH}{mm}. \end{aligned} \tag{5.33}$$

for 3mm bond wire $L_{Mutual} \cong 2.21\ nH$.

For a parallel IO scheme this would lead to the following reduction of bondwire inductance,

$$L_{Eff.} = \frac{(L_{Self} + L_{Mutual})}{2} \cong 2.77 nH, \tag{5.34}$$

with 17% reduction of bond wire inductance, which can be doubled, if an antiparallel IO scheme is applied:

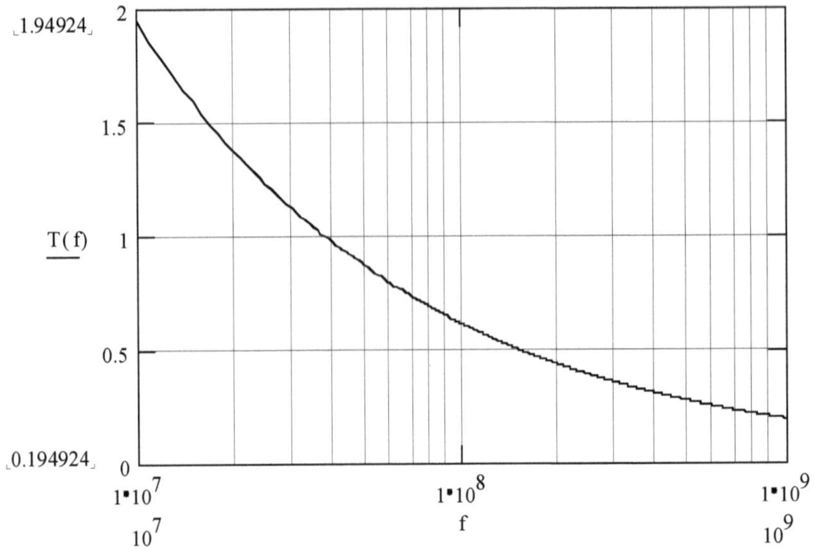

Fig. 5.11: Skin depth over frequency in case of this folding A/D converter design

(5.35) $$L_{Eff.} = 2L_{Self} - 2L_{Mutual} \cong 2.24nH \ ,$$

with 33% reduction of bond wire inductance.

Since it is not possible to use multiple IOs for all Signals in a SoC, the antiparallel IO scheme should be applied at least for the power supply network. As it is shown above, the magnitude of power bounces can be reduced at least by 1/3, without any additional measures, if neighboring VDD and GND connection are used.

5.3.3.2 Backside Connection

Another approach to reduce substrate noise is to provide a low impedance backside contact of the chip. This can be done by providing an extra metallization on the highly conductive substrate. This approach not only reduces the resistance of the substrate contact to ground but also reduces the inductance, since there is no bondwire required for this ground connection.

Using this approach, a 75% reduction of crosstalk coupling into sensitive nodes is possible [112]. The results found in the open literature indicate that this method provides the most effective reduction of the substrate crosstalk in heavily doped substrates for frequencies up to the 100 MHz range. At higher frequencies the skin effect in the highly doped substrate become dominant and cut off the low impedance backside connection.

Hence, the bulk material must be thinned to be less than one skin depth for the entire problem frequency range to make the backside contact an effective ground plane. At one skin depth from the surface of a conductor the current density has dropped to 1/e of its value on the surface:

$$T_{Skin}\left[mm\right]=10\cdot\sqrt{\frac{\rho}{\pi\mu f}}=10\cdot\sqrt{\frac{\rho}{4\pi^{2}\cdot 10^{-9}\cdot f}}\,. \tag{5.36}$$

with ρ = resistivity in [ohm cm] and f = frequency of interest.

The higher the frequency or the lower the resistivity the lower the skin depth which means that the wafer must be more and more thinned until it becomes unmanageably fragile at about 200 microns. It is important to keep in mind that a good square wave carries the first several odd frequency harmonics of the fundamental.

Thus, in the case of this folding A/D converter design with: $\rho=15\ \left[\mathrm{mOhm}\cdot\mathrm{cm}\right]$, T_{Skin} over the frequency axis looks like in Fig. 5.11. T_{Skin} for the fundamental and the odd harmonics in values taken from Fig. 5.11 are listed in Table 5.2.

A standard thickness for wafer to be backside metallized is 0.38mm. This is enough to reduce substrate noise coupled by the fifth harmonic of the fundamental clock frequency.

In this context the horizontal or lateral and vertical resistance are described by:

Table 5.2: Extraction of Fig. 5.11

f	T_{Skin}
40 MHz, fundamental	0.975mm
120 MHz, third (odd) harmonic	0.563mm
200 MHz, fifth (odd) harmonic	0.436mm
280 MHz, seventh (odd) harmonic	0.368mm

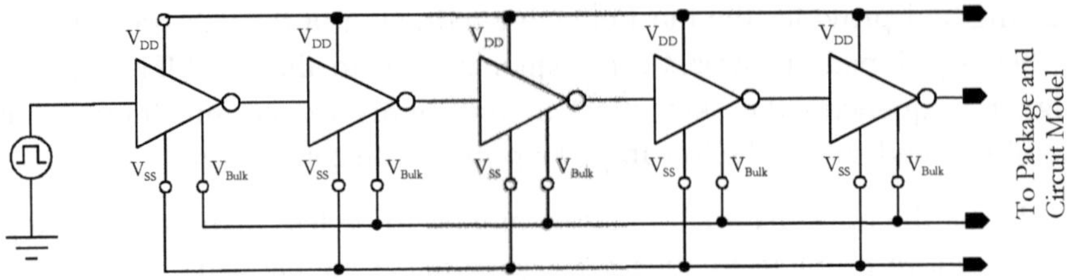

Fig. 5.12: Inverter chain switching model

$$R_{Lat} = \rho \cdot \frac{L}{W \cdot t} \tag{5.37}$$

and

$$R_{Vert} = \rho \cdot \frac{t}{W \cdot L} \,. \tag{5.38}$$

whereas $L =$ length, $W =$ width and $t =$ thickness of wafer.

As the substrate resistivity increases, the effect of surface point contacts causes the (lateral) resistance to increase and the current to be concentrated near the silicon surface. The thinner the wafer, the higher the lateral substrate resistance and the lower the vertical resistance.

In the case of this folding A/D converter design: The lateral resistance is about 0.41 Ohm/square; the vertical resistance is about 5.91 mOhm.

5.3.4 Digital Substrate Noise Generation

The performance of analog circuits in mixed-mode ICs is profoundly affected by coupled digital noise. This coupling can be caused by three mechanisms [113]:

- di/dt noise coupling from digital power supply,
- dv/dt coupling from switching source/drain nodes, and
- Impact ionization in the MOSFET channel.

The first mechanism introduces di/dt noise and resistive voltage drops due to inductance and resistance of the power-supply connections to the chip. Combination of this inductance and on-chip capacitance will also cause ringing of the power supply voltage owing to resonance effects. Due to the large amount of ground contacts to the substrate in every digital gate, the resistance between digital ground and substrate is very low and the ground noise and ringing is also present on the substrate.

Since the di/dt noise is dominated by switching currents of the digital circuitry, its magnitude is proportional to the value of the digital supply voltage. Hence the di/dt noise performance benefits directly from the supply voltage reduction in low power digital cores.

A risk in SoC design is, that designers may only care about schematic functionality and do not take parasitic effects into account. In the case of di/dt noise coupling, no effects are to to be seen in simulation if only ordinary block level simulations are performed. To prevent negative interaction of digital portions with the analogue functional blocks, it is strongly recommended to perform also simulations of the digital blocks with a appropriate package model. Due to the circuit complexity of large digital functional blocks, this can not be done by analogue simulations only. Hence a switching noise model should be added to the simulation of the analogue blocks.

As it is already stated in [114] two different kind of switching noise models can be introduced to the simulation environment, an inverter chain, which is shown in Fig. 5.12, or a multiplied inverter model, shown in Fig. 5.13.

The inverter chain cannot be called a realistic model of SoC digital circuit. Nevertheless it exhibit the right distribution of all frequency components, which can be present on the power supply or in the substrate. Therefore, the inverter chain can be called a good vehicle for maximization of digital noise rejection. Also the resonances based on RLC of the package model can be tuned towards the optimum noise rejection by simulating with this model.

Since the noise of this model is strongly dependent on the delay of the inverter, this model is especially suitable for modeling asynchrounus digital circuit noise.

A more realistic model for the transient behavior of synchronous digital circuit noise is the multiplied inverter switching model, also introduced in [114].

To achieve a realistic model, the number of simultaneously switching digital circuitry has to be extracted and converted into an equivalent multiplier of the inverter transistors, M_N for the NMOS and M_P for the PMOS. The switched parasitic capacitances of the transistors are C_N and C_P, whereas the capacitor C_{Load} represents equivalent capacitive load of all comprised digital cells. To achieve a realistic transient behavior, this extraction should be done

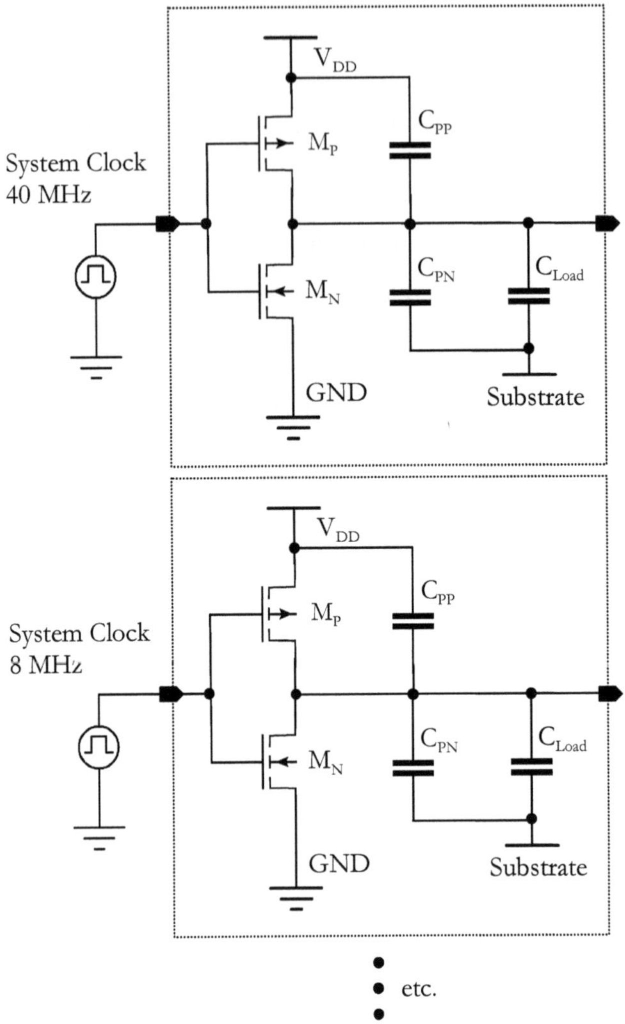

Fig. 5.13: Multiplied inverter switching model

separately for all synchronous switching digital circuitry with the same clock frequency. This model gives a much more realistic transient behavior of the digital noise. Maximum disturbances at the power supply and the substrate node can be determined and the timing of analogue functional block can be tuned towards timings with a minimum signal distortion.

Hence it is much more suitable for timing optimization than the inverter chain model.

Since the digital noise of the multiplied inverter model is synchronous to the clock signal, this model is more suitable to determine the effect of synchronous digital noise on the functionality of an IC.

The mechanism of di/dt noise is often the dominant cause of substrate noise, especially if interconnect inductance and substrate biasing schemes are poorly designed.

The resulting substrate voltage waveform of the dv/dt noise are voltage spikes at the low resistance substrate node, located at the clock edges. This is in contrast to noise coupling from power supply, where switching digital gates cause an RLC ringing of the substrate node voltage. Hence both effects can be differed by their shape.

The amount of dv/dt digital circuit noise is strongly dependant on two parameters:

- The impedance of the substrate bias (N-well, P-well)
- The parasitic capacitance to the substrate

To prevent digital noise from penetrating the substrate, all digital wells should be biased using low ohmic connections. Therefore, substrate contacts should be used as often as possible in digital cells.

It is also recommended that the bias voltage used for biasing the wells does not carry any transient current, since this would cause ringing RLC ringing. This ringing will inject more noise to the substrate then noise will be rejected by the low ohmic substrate connection.

The effect of dv/dt noise, caused by switched drain/source nodes of the digital circuitry, can also be modeled by the inverter chain shown in Fig. 5.12 or the multiplied inverter shown in Fig. 5.13, whereas the effect of imperfect substrate biasing can be modeled by a suitable substrate model.

Digital dv/dt noise can also be injected by parasitic capacitances between switching networks and the substrate. Preferably this is caused by clock distrubution networks or huge digital buses with heavily switching activities. As already stated in the case of capacitive noise coupling, the first step is to choose the right metal layer. To prevent substrate noise, an upper metal layer should be chosen. In addition a substrate shielding can be used by terms of intermediate metal layers. Since this normally leads to an inacceptable increase of signal run time, isolation from digital signals to the substrate is noramally done by reversed biased wells underneath the bus or clock network. Of cause this well should be also biased with low impedance connections to V_{DD} or V_{SS} to ensure proper shielding and prevent additional noise injection.

The effect of ionization in the MOSFET channel turned out to be negligible for the technology used for fabrication of this folding A/D converter design.

From simulations shown in [113] it can be seen from the signal characteristic that for an extremely low inductance in the power lines the substrate noise is generated mainly by capacitive coupling from source/drain nodes. This can only be reduced by increasing the number of substrate contacts to a quiet digital ground when increasing the inductance, the capacitively coupled noise from the source/drain nodes becomes less important and the substrate noise becomes dominated by noise coupling from the power supply lines which can be seen from higher signal peaks instead of signal edges. For large inductances, the substrate noise is dominated by ringing of the damped LC resonance circuit, formed by the power line inductance and the decoupling on-chip capacitance. Any resistance in series with the capacitor will cause damping of the LC resonance circuit and affect the ringing.

To conclude: Only if the power supply interconnect inductance is very low and the power supply noise is not dominant anymore, a further reduction of substrate noise is on chip possible by increasing the number of substrate contacts and using a dedicated well and substrate bias with low connection impedance.

Another measure is to reduce the effects of troublemakers (e.g. switching of digital output drivers, digital supply voltage) and their impacts on the substrate. Possibilities are: use of low digital supply voltage, use of slow rise and fall times for digital transients, staggering the timing of on-chip drivers, and switching off digital parts not in use.

In any case the use of suitable simulation models for the package and the digital circuitry is essential for a good estimation, whether an improvement of embedded analogue functional performances is within reach or not, for the project given boundary conditions.

5.3.5 Substrate Coupling

As shown above, the chip substrate acts as a collector and distributor of noise in the IC. Essentially every chip carries voltage transients on chip signal wires and I/O pads and power supply rails are capacitive coupled to the chip substrate. This includes energy from PCB reflections back to I/O pads and transmitted to the substrate via I/O protect devices. Peak voltages of the coupled noise are frequency dependent due to constructive and destructive interference of the various frequencies.

Highly doped p+ bulk wafers with grown epitaxial layers are easiest to analyze when the p+ bulk can be seen as a pseudo ground plane under the epitaxial layer and thus can be used as a single node. This is the case, if the p+ bulk is degeneratively doped (about 0.006 Ohm cm) and the thickness of the substrate is below the skin depth for the frequencies of interest (see also chapter 5.3.3.2). This can be approximately reached with a low inductance and low resistance connection to the entire chip backside using a titanium-gold metallization (see 5.3.3.2).

Due to the low resistance bulk substrate, separation of noise transmitters to noise receivers makes almost no difference in coupled noise. Most current flows via the epi region vertically and renders n well guard-rings basically ineffective in shielding a device from other than its nearest neighbor.

Experimental and simulation results [108] for heavily doped bulk indicate that analog and digital circuits should be separated by at least four times the thickness of the lightly doped epitaxial layer. Beyond this distance, further separation does not significantly reduce substrate crosstalk.

p+ guard rings, when spaced close to the epi thickness away and tied to a non-contaminated (i.e. noiseless) ground can substantially reduce substrate coupled noise with respect to an NMOS device (this is best achieved by connecting to a dedicated package pin). The guard ring must be located closely enough to decrease the horizontal resistance between sensitive node and p+ guard ring in order to decrease the vertical resistance between sensitive node and noisy substrate. If a non-dedicated contact is used for the p+ guard ring (connected to all substrate contacts on the chip), the vertical resistance between substrate contact diffusion and noisy substrate becomes lower (because of the large total size of the diffusion) in comparison to the horizontal resistance and the substrate coupled noise to the sensitive node increases.

A second capability of guard rings is to equalize noise coupling to matched devices by coupling it in common mode. This helps in the case when differential signal path is used for construction of analog circuits (these circuits reject common-mode signals). An n+ contact surrounding the perimeter of the n-well is also helpful in distributing the coupling to the n-well and make the coupling almost completely common mode for both devices. In addition, guard rings help to minimize body effect and capacitive coupling. The p+ guard ring helps to equalize the potential of the substrate around an NMOS device and reduce body effects as well as bulk-to-source coupling if the NMOS device is referenced to the same ground that is tied to the p+ substrate guard ring.

To the first order, the coupling from the substrate to a circuit node is proportional to the capacitance to substrate from that node and the impedance of the coupled node. As a result, substrate noise coupled to PMOS devices located in an n-well tied to low inductance power supply line will be lower than for an NMOS device located in the epitaxial layer.

5.4 Summary of Recommended Design Practices for Low-Power, High-Speed CMOS Analog-to-Digital Converter for Embedded Systems

This is to sum up all design practices for low substrate crosstalk.

1. Reduce the noise generation ("quiet the perpetrator")
   a. Use a package with the lowest I/O inductance possible for your project boundary conditions.
   b. Use multiple antiparallel power supply I/Os to reduce the I/O inductance and the resulting power bounces.
   c. Use package model for simulations to prevent resonance of the I/O impedance.
   d. If it is possible, make use of low voltage digital cores.
   e. Optimize the switching transients by tuning the rise and fall times.
   f. Reduce simultaneous switching activity by timing optimization especially in terms of I/O drivers.
   g. Lower the switching function bus impedance.
   h. Reduce the switching activity by switching off all digital parts not in use.
   i. Sample the analogue signals at times of minimum noise values by timing optimized sample & hold stages.
   j. Use a digital circuit model for simulation to optimize switching activities well as analogue and digital timing
   k. Isolate sensitive circuits from noisy signals ("isolate the victim")

2. Maximize the impedance form noise source to noise victim. In terms of low ohmic substrate this is only possible for the lateral direction by the use of well guard rings connected to dedicated I/O pins.
   a. Minimize the impedance of the chip substrate node, e.g. by use of low inductance packages, multiple substrate connections and backside side contacts where possible.
   b. Make use of on-chip shielding by use of intermediate metal shielding for shielding of sensitive signals or dedicated wells to shield the substrate form digital noise

3. Make the analogue functional block more noise tolerant ("immunize the victim")
   a. Make use of designs with high power supply rejection ratio and common mode rejection ratio.
   b. Make use of differential signal path circuit topology. This is to cancel the influence of noise coupling on common nodes.
   c. Use the analogue functional blocks with a minimum frequency bandwidth, to filter high frequency noise.
   d. Keep internal nodes at a minimum impedance to drain the noise a quick as possible. This is especially important for reference voltage networks.
   e. Try to optimize analogue signal sampling by timing optimization. This should be done be the use of digital noise models in the analogue simulation environment.
   f. Make use of common-centroid geometry layout techniques to equalize the influence of substrate noise.
   g. Verify your physical design in terms of signal immunity by the use of parasitics extracted simulations.

CHAPTER 6

EVALUATION AND MEASUREMENTS

Epitome

In this chapter evaluation and measurements of the integrated 10-bit folding and interpolating A/D converter are presented. Generally speaking, there are two different methods to test high-speed A/D converters. The first one is a purely analog method. The other one is a digital method that uses high-speed memory and digital signal processor (DSP). The following section will discuss both methods.

6.1 Methodology for Analog Measurement

Analog measurement of A/D converters relies on analysis of the analog signal reconstructed from the digital output of the A/D converter under test. For this purpose a high-performance D/A converter is connected to the outputs of the A/D converter. If the input signal of the A/D converter is a pure sine wave, as illustrated in Fig. 6.1, this arrangement yields a reconstructed sine wave at the output of the D/A converter.

The reconstructed sine wave is fed into a spectrum analyzer, which carries out the *fast Fourier transform* (*FFT*). Based on the FFT, the *signal-to-noise ratio* (*SNR*) can be determined and the resolution can be calculated as *effective number of bits* (*ENOB*) as Eq. (2.27).

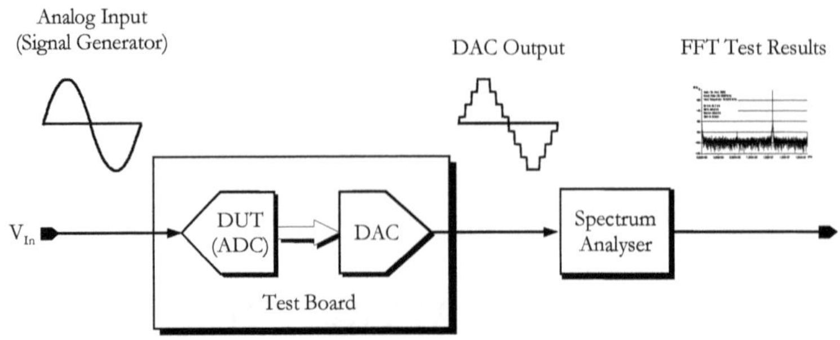

Fig. 6.1: Block diagram of the analog measurement

There is one good rule of thumb to follow when selecting components for testing A/D converters: the component accuracy should be about five to ten times higher than the accuracy of the tested A/D converter [13]. That is to say, when testing an A/D converter the test system should have accuracy of 2 to 3 bits higher than the A/D converter under test. The ideal SNR of a 10-bit A/D converter is $61.96\ \mathrm{dB}$. The SNR of the test system should thus exhibit the accuracy at least of 12-bit or $74\ \mathrm{dB}$.

Thus the resolution of the D/A converter performing the sine wave reconstruction in Fig. 6.1 must be at least about 2~3 bits better than the A/D converter under test, as discussed above.

The disadvantage of the analog measurement is that the digitized output signals of the A/D converter are not saved in binary form. This leads to great problems, if the output signals of the A/D converter have to be post-processed.

6.2 Methodology for Digital Measurement

Another method to measure the performance of the A/D converter is to employ digital signal processing for evaluation. The corresponding arrangement is illustrated in Fig. 6.2.

In this case, the output signals of the A/D converter (*DUT*, device under test) are first fed into a first-in first-out (*FIFO*) memory and then the saved data are processed by a DSP, which, for example, can perform digital signal processing such as digital filtering or FFT preprocessing. The FFT analysis of the processed data is carried out using a microprocessor.

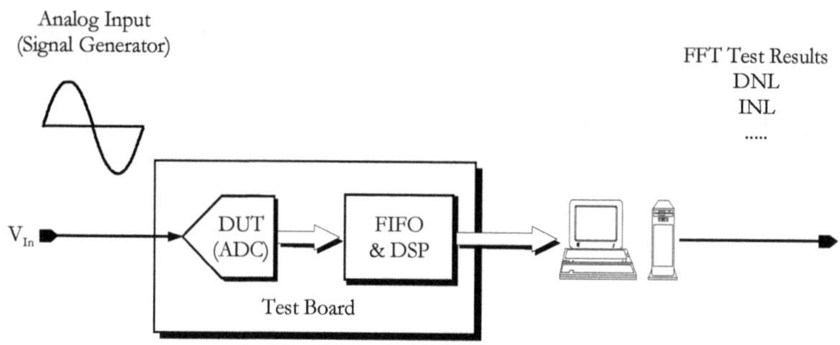

Fig. 6.2: Block diagram of the digital measurement

A 10-bit parallel A/D converter running at 50 MHz, for example, generates a data flow of 500 Mbit/sec. The data transfer rate of each data line is thus 50 Mbit/sec. Since the data line has its physical transfer bottleneck, the digital measurement is limited by the data transfer rate of the data line. Moreover, the digital measurement is also constrained by the reading speed of the FIFO memory and the memory size limits the length of the measuring period.

The analog method is suitable for the real-time test, but it gives only result of the FFT analysis. On the other side, since the digital method saves output data of the A/D converter for further investigation, it is more suitable for debugging and other analyses such as DNL and INL.

For the test of the implemented 10-bit folding and interpolating A/D converter, a digital measurement test-bed has been implemented in this work. The FIFO memory has 18 parallel channels and the memory capacity is up to 64 k words. The FIFO is controlled by a microprocessor using the LabView software. For this purpose a program in G-programming language has been written that runs under LabView. This program also computes the FFT and evaluates all performance of the A/D converter under test.

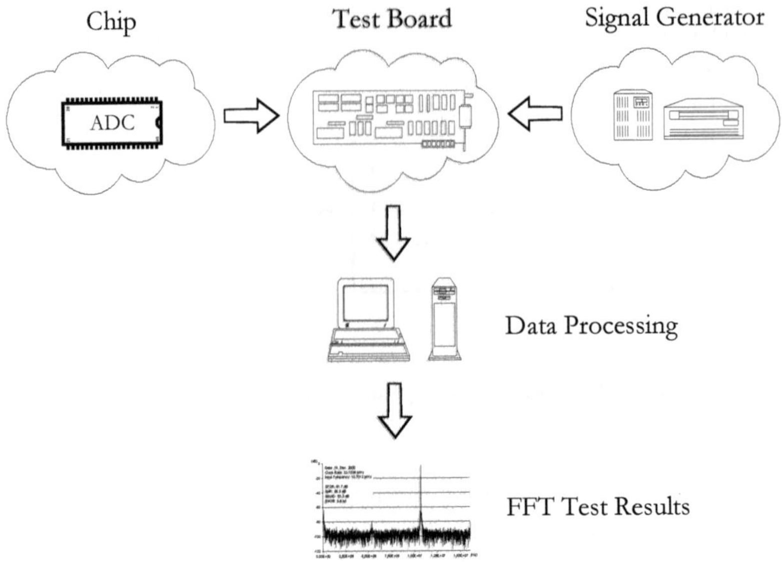

Fig. 6.3: FFT analysis

6.3 Measurement Results

6.3.1 Introduction

The measurement of A/D converter can be affected by numerous noise sources interfering with the measurement. To explain possible noise sources in the test environment, Fig. 6.2 has been extended to Fig. 6.3 to indicate the measurement set-up more in detail. The FFT analysis is performed on the data of the tested A/D converter and is thus affected by all imperfections of the converter measured. But, unfortunately, it is also affected by imperfections of the test environment. Thus, e.g. considering noise, noise is not only generated by the A/D converter under test, but also by the test board and the signal generator, as depicted in Fig. 6.3.

As shown above (see Eq. (2.26)), the SNR of an ideal 10-bit A/D converter is 61.96 dB, which means the *root-mean-square* (*RMS*) quantization noise level should be -61.96 dB below the full scale converter signal level. But the value of single bin of the FFT plot depends on the number of regarded records. Fig. 6.4 shows the RMS quantization noise level and the FFT floor for

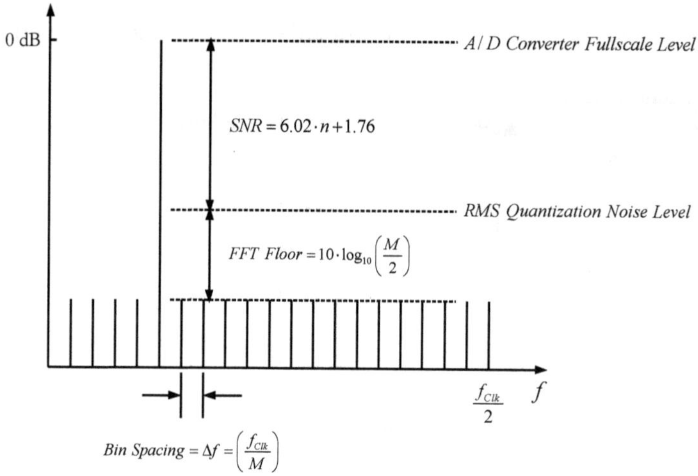

Fig. 6.4: Theoretical FFT noise floor

a A/D converter full-scale signal level at 0 dB. *M* indicates the number of regarded records of the conversion [15].

In order to be able to test a 10-bit A/D converter properly, all harmonic frequencies of the sine wave signal generator should be lower than the FFT floor shown in Fig. 6.4. Providing that $M = 8192$, the harmonic frequencies should be 98.08 dB below the ADC full-scale level to ensure that their contribution is not larger than the quantization error. Since commercial sine wave signal generators usually provide harmonic distortion at about -40 dBc, their *total harmonic distortion* (*THD*) must be improved by additional measures, e.g. by additional *bandpass* or *lowpass filtering* (*BPF* or *LPF*) of the output signal of the generator. Apart from the sine-wave signal generator, the clock generator may introduce clock jitter and phase noise which can also influence the measurement unless a proper generator is selected.

Concerning the implementation of the test board, the crosstalk is the main problem. The crosstalk arises mainly due to coupling between analog and digital signal interconnects and analog and digital power supply wires including ground. The crosstalk generates disturbances, which raise the noise floor of the output signal of the FFT and might even exceed the noise of the converter itself.

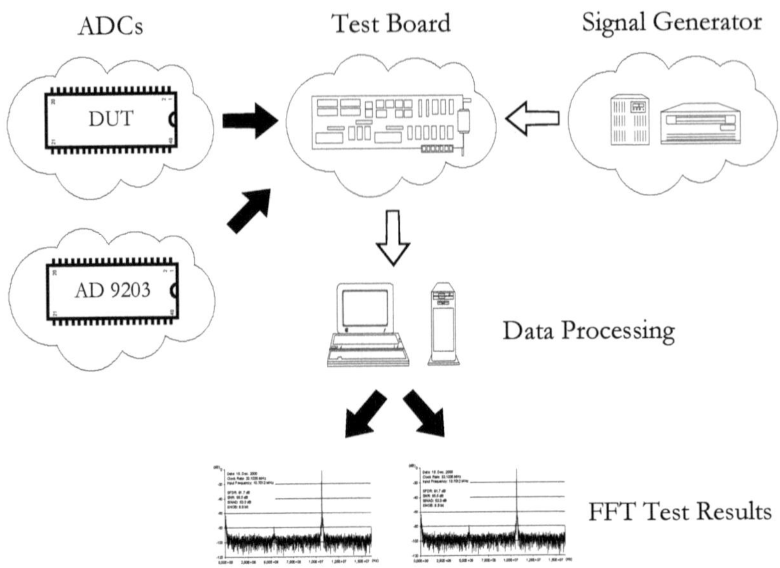

Fig. 6.5: Comparison measurements between the implemented folding and interpolating ADC and the commercial ADC AD9203

In order to be able to verify the performance of the measurement set-up and to be able to evaluate the implemented 10-bit folding and interpolating A/D converter, a commercial A/D converter made by Analog Devices Inc. (type AD9203) has been tested under the same conditions. The test procedure is illustrated in Fig. 6.5 and the test results and some main characteristics of AD9203 are listed in Appendix C.

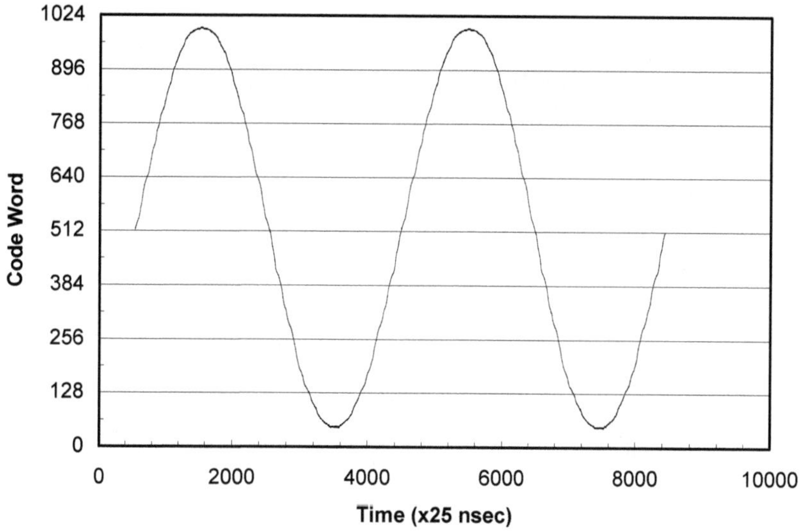

Fig. 6.6: Oversampling measurement of the implemented A/D converter

6.3.2 Oversampling Measurement

The oversampling measurement serves to find codes missing during the conversion. The results of this measurement are shown in Fig. 6.6, which depicts a reconstructed sine wave signal. The sampling rate was 40 MSPS, equivalent to 25 nsec, and the input frequency of the sine wave was 10.4 kHz. The corresponding *oversampling rate* (*OSR*) equals to 3846.15.

This measurement was carried out without a bandpass filter.

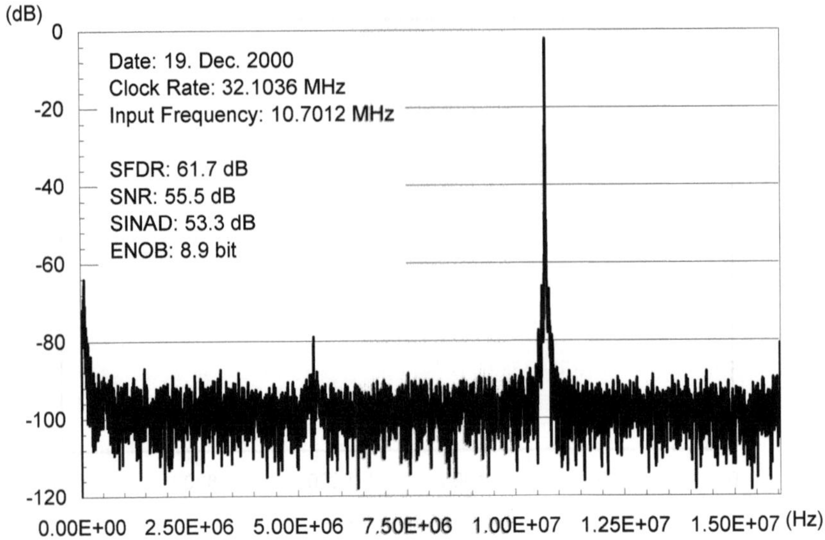

Fig. 6.7: Measurement of sampling rate at 32.1036 MSPS

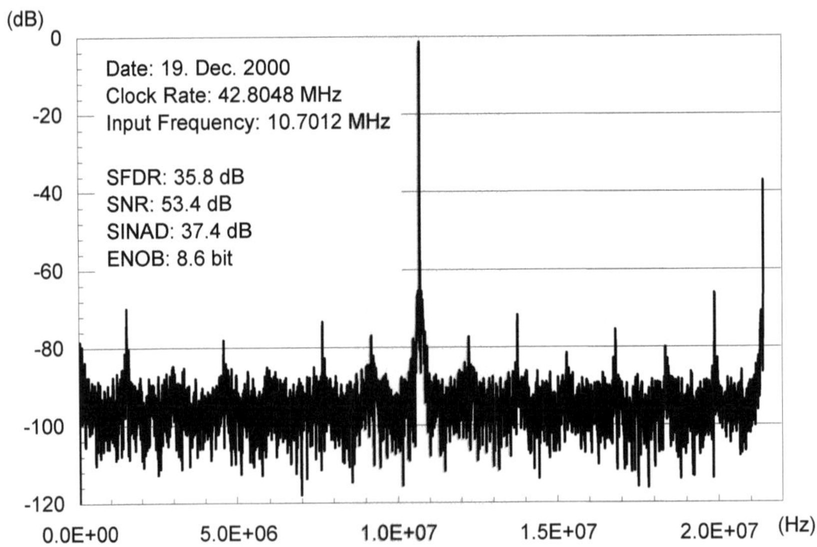

Fig. 6.8: Measurement of sampling rate at 42.8048 MSPS

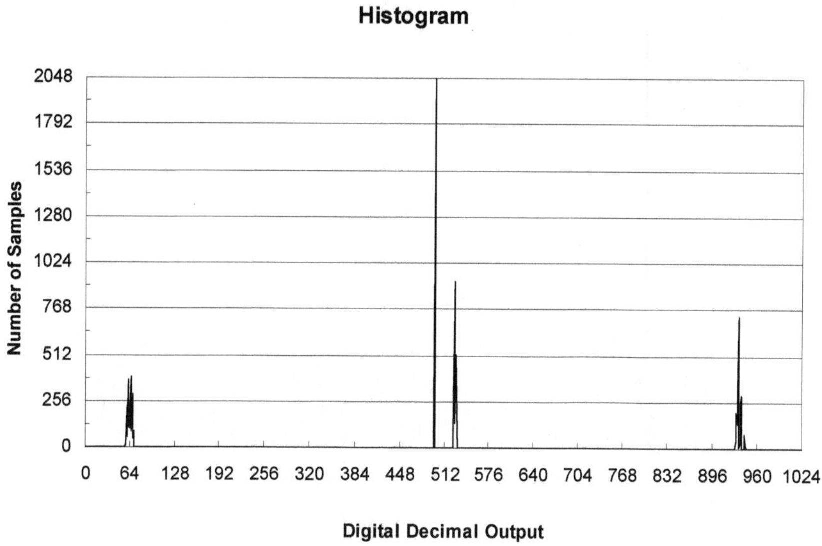

Fig. 6.9: Statistical observation of the clock jitter

6.3.3 FFT Analysis

A bandpass filter (*BPF*) with centered frequency of 10.7012 MHz has been used to reduce the noise level and to improve the THD of the signal generator. Consequently, according to the Fig. 6.3, the most of the noise comes only from the test chip and the test board.

Fig. 6.7 shows the results of an FFT analysis. The input frequency is 10.7012 MHz and the sampling rate is 32.1036 MSPS. The measured spurious free dynamic range (*SFDR*) is 61.7 dB and the corresponding SNR and signal-to-noise and distortion ratio (*SINAD*, also abbreviated as *SNDR*) are 55.5 dB and 53.3 dB, respectively. The ENOB under these conditions is 8.9 bit.

Fig. 6.8 shows another FFT analysis. The input frequency is still 10.7012 MHz which is also the center frequency of the bandpass filter, but the sampling rate has been increased to 42.8048 MSPS. The measured SFDR is 35.8 dB and the corresponding SNR and SINAD are 53.4 dB and 37.4 dB, respectively. The ENOB is now 8.6 bit.

In the latter case, the SFDR has decreased drastically down to 25.9 dB because the 2nd harmonic frequency in this case is quite high. The same phe-

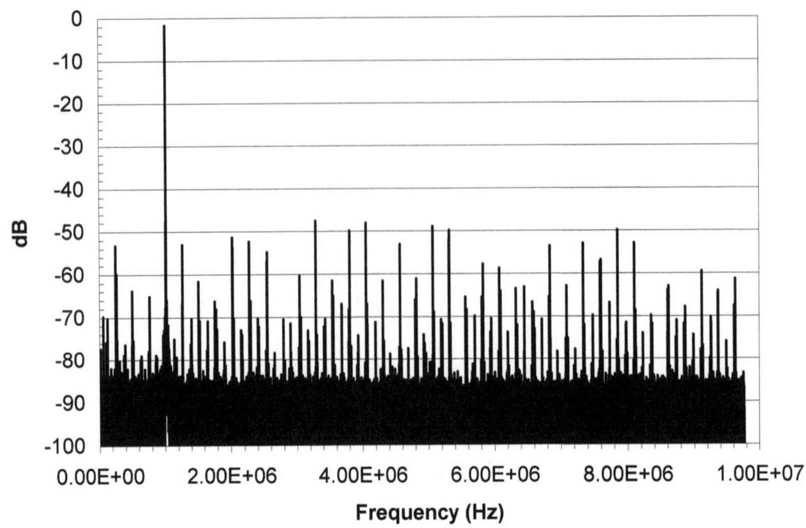

Fig. 6.10: FFT analysis for an input frequency of 1.07178 MHz

nomenon can be also found in SINAD. The decrease amounts to 15.9 dB. In contrast, the SNR has decreased only by 2.1 dB since the first 5 harmonic frequencies and the DC contribution are not considered in the SNR computation. The different behavior between SINAD and SNR shows that the main reason for the degrading of the performance is from the first 5 harmonics and the DC part.

Fig. 6.9 shows the histogram of the same measurement as illustrated in Fig. 6.8. There are 8192 data records in total. Since the sampling frequency is just four times the input frequency, theoretically there shall be only four values. It can be seen in the plot that the sampled data are distributed into four groups. One of them has reached its designed value, 2048, because the sine wave signal generator is triggered by the clock signal and on the other hand, the converter also meets its goal. The other data for the remaining groups are distributed within a certain range. This suggests that there are problems with clock jitter, stemming either from the converter or from the test environment.

As a comparison basis, standard high-speed A/D converters are usually measured with an input frequency around 1 MHz. Fig. 6.10 illustrates this kind of FFT analysis, which was performed for an input frequency of 1.07178 MHz as input signal and sampled at 20.0 MSPS. The plot shows that the noise floor is at about 85 dB, but the amount of harmonics is very high. The measured SFDR is 47 dB. This measurement has been carried out without the bandpass filter.

6.4 Summary

There are basically two ways to measure the performance of the A/D converters. The first one is analog measurement: it employs a high performance D/A converter to convert the signal digitized by the A/D converter under test back to the analog signal. This enables dynamic analysis of the reconstructed analog signal to be carried out using the spectrum analyzer. This treatment is suitable for high-speed solution, but the result is more a qualitative analysis rather than a quantitative analysis. Another possibility is a digital measurement by saving the digitized data in an external memory. The saved data can be digitally processed for further analysis in detail, if necessary. Only the memory size and speed limit the measuring period and the conversion speed of the ADC. As a consequence, the method is suitable for quantitative analysis at lower conversion rates.

The measured results of the implemented folding and interpolating converter have been presented in this chapter. The oversampling measurement shows the capability to reconstruct an input sine wave at the sampling rate of 40 MSPS. The FFT analysis shown in Fig. 6.7 and Fig. 6.8 shows the behavior of the A/D converter at sampling frequencies of 32.1036 MSPS and 42.8048 MSPS, respectively. Fig. 6.9 illustrates the phenomenon of the clock jitter at high input frequency of Fig. 6.8. Fig. 6.10 shows the FFT analysis for an input frequency of 1.07178 MHz at sampling rate of 20 MSPS.

The current version of the folding and interpolating A/D converter does not meet the expected accuracy of 10-bit yet. The measurements shown here suggest that the reasons for this is not noise but the clock jitter.

CHAPTER 7

GENERAL CONCLUSIONS AND OUTLOOK

This book is devoted to high-speed CMOS analog-to-digital converters that feature low power dissipation and small die area. The principle employed is based on folding and interpolating A/D converter architecture and it is especially convenient for embedding into large systems.

A concept has been presented that allows a modular design of such converters. This was supported by component development and architecture design and optimization. A 10-bit prototype has been integrated and tested. This particular converter is intended for use in digital beamformers for ultrasound equipments [115]. Due to its low-power and high-speed properties, a low-cost handheld ultrasound unit is realizable. The unit will contain 4 channel-A/D converters and beamformers in the coming future. The final product should have even 8 ~ 16 channel-A/D converters and corresponding beamformers.

Due to the modularity of this type of A/D converter, resolution scaling, e.g. in the range 8 ~ 12-bit, is possible. Formulae have been developed that enable prediction of die size and power dissipation before the design work starts. Quality of such prediction can be improved by methods like "learning curve", i.e. the converter parameters of every implementation are measured, characterized, and normalized. The architecture relies on scalable modules, which work as "LEGO bricks" to build A/D converter of other architectures or other resolutions. Accurate prediction can be obtained using averaged scaling units from various implementations.

The co-integration of analog circuits together with the digital signal processing generates crosstalk between the analog and the digital signals. This poses severe problems, particularly for A/D converters because it may lead to reduction of A/D converter resolution. But, the idea of the concept of "system-on-a-chip" definitely requires co-integration of A/D and D/A converters, if both, analog and digital signal are to be processed on a single chip. Hence, investigations of crosstalk are of importance for a robust design. Beside the simulation efforts that must be taken into consideration, careful layout and characterization of parasitic components are also essential issues of high-speed analog circuits design. Future work in the area of low-power high-speed CMOS A/D converter design should pay increased attention to these topics.

APPENDIX A

ANALYSIS FOR SMALL SIGNAL OPERATION

Differential Amplifier with Active Load

Assuming that both transistor pairs in Fig. A.1, M1 and M2, M3 and M4, are with the same dimensions, then g_L, g_m, and g_{DS} can be written as [3]:

$$g_L = g_{m3} + g_{DS3} = g_{m4} + g_{DS4} , \tag{A.1}$$

$$g_m = g_{m1} = g_{m2} , \tag{A.2}$$

$$g_{DS} = g_{DS1} = g_{DS2} , \tag{A.3}$$

and

$$g_L >> g_{DS} . \tag{A.4}$$

From the nodal analysis, three relationships concerning v_A, v_B, and v_C can be obtained:

$$v_A \cdot g_L + (v_A - v_C) \cdot g_{DS} + g_m \cdot (v_1 - v_C) = 0 , \tag{A.5}$$

$$v_B \cdot g_L + (v_B - v_C) \cdot g_{DS} + g_m \cdot (v_2 - v_C) = 0 , \tag{A.6}$$

and

$$\begin{aligned} &(v_C - v_A) \cdot g_{DS} - g_m \cdot (v_1 - v_C) + v_C \cdot g + \\ &(v_C - v_B) \cdot g_{DS} - g_m \cdot (v_2 - v_C) = 0 . \end{aligned} \tag{A.7}$$

Rearranging the equations, (A.5), (A.6), and (A.7), leads:

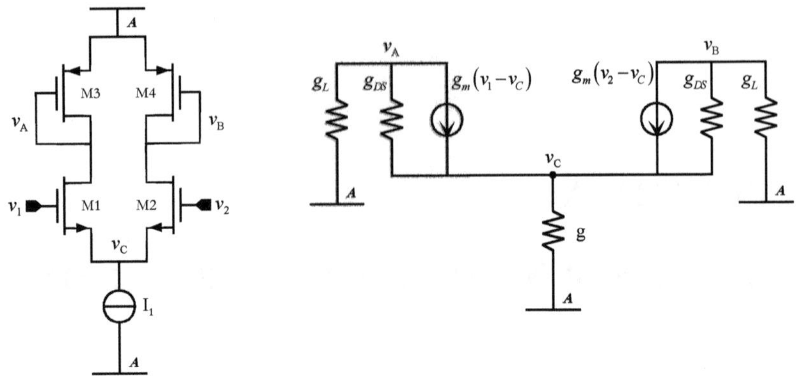

Fig. A.1: Differential CMOS amplifier with PMOS active load

$$v_A = \frac{(g_{DS}+g_m)\cdot v_C - g_m\cdot v_1}{(g_L+g_{DS})}, \tag{A.8}$$

$$v_B = \frac{(g_{DS}+g_m)\cdot v_C - g_m\cdot v_2}{(g_L+g_{DS})}, \tag{A.9}$$

and

$$v_C = \frac{g_m\cdot(v_1+v_2)+g_{DS}\cdot(v_A+v_B)}{(g+2g_m+2g_{DS})}. \tag{A.10}$$

Adding Eqs. (A.8) and (A.9) leads:

$$v_A+v_B = \frac{2\cdot(g_{DS}+g_m)\cdot v_C - g_m\cdot(v_1+v_2)}{(g_L+g_{DS})}. \tag{A.11}$$

Substituting Eq. (A.11) into Eq. (A.10) gets:

$$\begin{aligned} v_C &= \frac{g_L\cdot g_m}{2g_L\cdot(g_{DS}+g_m)+g\cdot(g_L+g_{DS})}\cdot(v_1+v_2) \\ &= f(v_1,v_2). \end{aligned} \tag{A.12}$$

Hence, the node voltage v_A and v_B can be written as:

$$v_A = -\frac{g_L g_m \left(g_{DS} + g_m\right)\left(v_1 - v_2\right) + g g_m \left(g_L + g_{DS}\right) v_1}{\left(g_L + g_{DS}\right)\left[2 g_L \left(g_{DS} + g_m\right) + g\left(g_L + g_{DS}\right)\right]}, \tag{A.13}$$

and

$$v_B = -\frac{g_L g_m (g_{DS} + g_m)(v_2 - v_1) + g g_m (g_L + g_{DS}) v_2}{(g_L + g_{DS})\left[2 g_L (g_{DS} + g_m) + g(g_L + g_{DS})\right]}. \tag{A.14}$$

The differential-mode voltage gain A_{DM} can be expressed as:

$$\begin{aligned} A_{DM} &= \frac{v_{Out,\,diff}}{v_{In,\,diff}} = \frac{v_B - v_A}{v_1 - v_2} \\ &= -\frac{g_m \left[2 g_L (g_{DS} + g_m) + g(g_L + g_{DS})\right]}{(g_L + g_{DS})\left[2 g_L (g_{DS} + g_m) + g(g_L + g_{DS})\right]} \\ &= -\frac{g_m}{(g_L + g_{DS})}. \end{aligned} \tag{A.15}$$

Further, supposing that $g = 0$ and for the diode-like connected MOS loads $g_{DS} << g_L$, the differential-mode voltage gain A_{DM} can be simplified as:

$$A_{DM} \cong -\frac{g_m}{g_L}. \tag{A.16}$$

APPENDIX B

SOURCE CODES FOR BINARY CODING

```
module decoder_1of64 (oneof, bincode);

    input {63:0} oneof;
    output {5:0} bincode;
//    reg {5:0} bincode;


// 66665555 55555544 44444444 33333333 33222222 22221111
11111100 00000000
// 32109876 54321098 76543210 98765432 10987654 32109876
54321098 76543210
// 01010101 01010101 01010101 01010101 01010101 01010101
01010101 01010101
   assign bincode{0} = |{oneof{ 0}, oneof{16}, oneof{32},
oneof{48},
                         oneof{ 2}, oneof{18}, oneof{34},
oneof{50},
                         oneof{ 4}, oneof{20}, oneof{36},
oneof{52},
                         oneof{ 6}, oneof{22}, oneof{38},
oneof{54},
                         oneof{ 8}, oneof{24}, oneof{40},
oneof{56},
                         oneof{10}, oneof{26}, oneof{42},
oneof{58},
                         oneof{12}, oneof{28}, oneof{44},
oneof{60},
                         oneof{14}, oneof{30}, oneof{46},
oneof{62}};


// 66665555 55555544 44444444 33333333 33222222 22221111
11111100 00000000
// 32109876 54321098 76543210 98765432 10987654 32109876
54321098 76543210
```

```
// 01100110 01100110 01100110 01100110 01100110 01100110
01100110 01100110
   assign bincode{1} = |{oneof{ 1}, oneof{17}, oneof{33},
oneof{49},
                    oneof{ 2}, oneof{18}, oneof{34},
oneof{50},
                    oneof{ 5}, oneof{21}, oneof{37},
oneof{53},
                    oneof{ 6}, oneof{22}, oneof{38},
oneof{54},
                    oneof{ 9}, oneof{25}, oneof{41},
oneof{57},
                    oneof{10}, oneof{26}, oneof{42},
oneof{58},
                    oneof{13}, oneof{29}, oneof{45},
oneof{61},
                    oneof{14}, oneof{30}, oneof{46},
oneof{62}};


// 66665555 55555544 44444444 33333333 33222222 22221111
11111100 00000000
// 32109876 54321098 76543210 98765432 10987654 32109876
54321098 76543210
// 01111000 01111000 01111000 01111000 01111000 01111000
01111000 01111000
   assign bincode{2} =
|{oneof{62:59},oneof{54:51},oneof{46:43},oneof{38:35},

oneof{30:27},oneof{22:19},oneof{14:11},oneof{ 6: 3}};

// 66665555 55555544 44444444 33333333 33222222 22221111
11111100 00000000
// 32109876 54321098 76543210 98765432 10987654 32109876
54321098 76543210

// 01111111 10000000 01111111 10000000 01111111 10000000
01111111 10000000
   assign bincode{3} = | {oneof {62:55}, oneof {46:39},
                     oneof {30:23}, oneof {14:7}};


// 01111111 11111111 10000000 00000000 01111111 11111111
10000000 00000000
   assign bincode{4} = | {oneof {62:47}, oneof {30:15}};


// 01111111 11111111 11111111 11111111 10000000 00000000
00000000 00000000
   assign bincode{5} = | oneof {62:31};
```

```
endmodule
```

APPENDIX C

TEST RESULTS: FOLDING ADC VS. AD9203

Following figures (Fig. C.1, Fig. C.2, Fig. C.3, and Fig. C.4) are measurement results of the implemented folding and interpolating A/D converter and the commercial A/D converter (type AD9203 from Analog Device, Inc.).

The measurement results are summarized in Table C.1.

Table C.1: Summaries of measurements

	Implemented ADC until now	Commercial ADC
DNL	-1 ~ 2 LSB	-0.4~0.3 LSB
SNR at about 10 kHz	35.9 dB	48.6 dB
SNR at about 1 MHz	35.0 dB	28.0 dB
Power dissipation	(60+40) mW at 3.3 V	74 mW at 3 V
Embedded systems	SoC possible	Only off-chip version

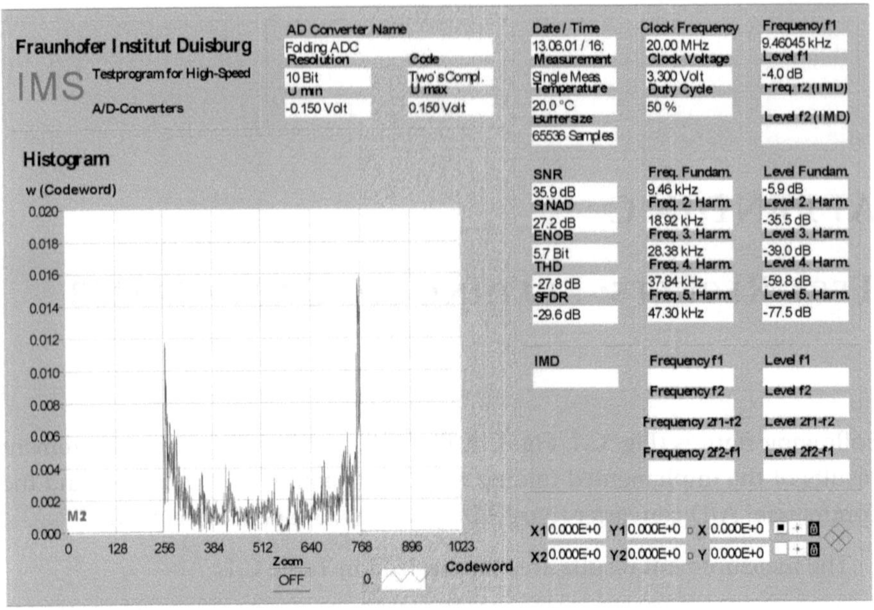

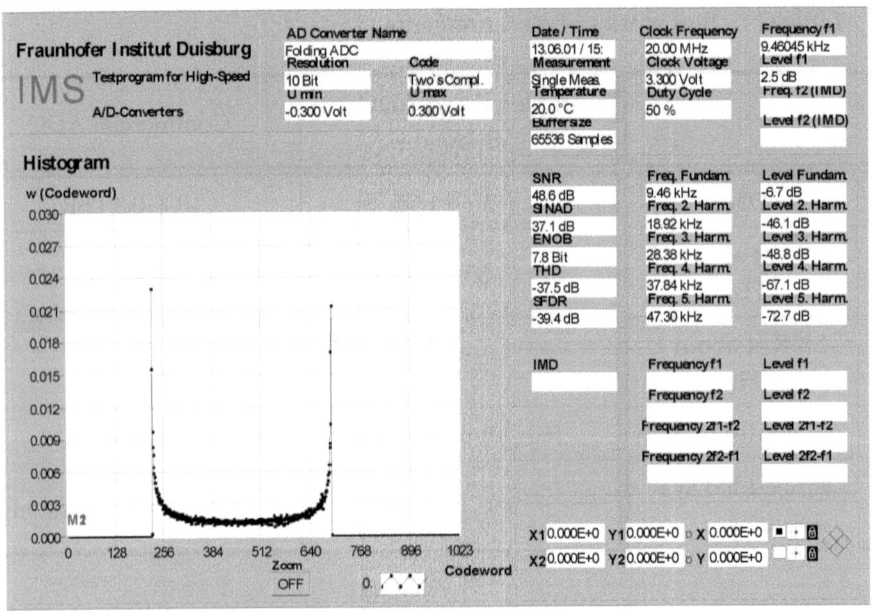

Fig. C.1: Histogram of the implemented ADC (above) and the commercial ADC (below)

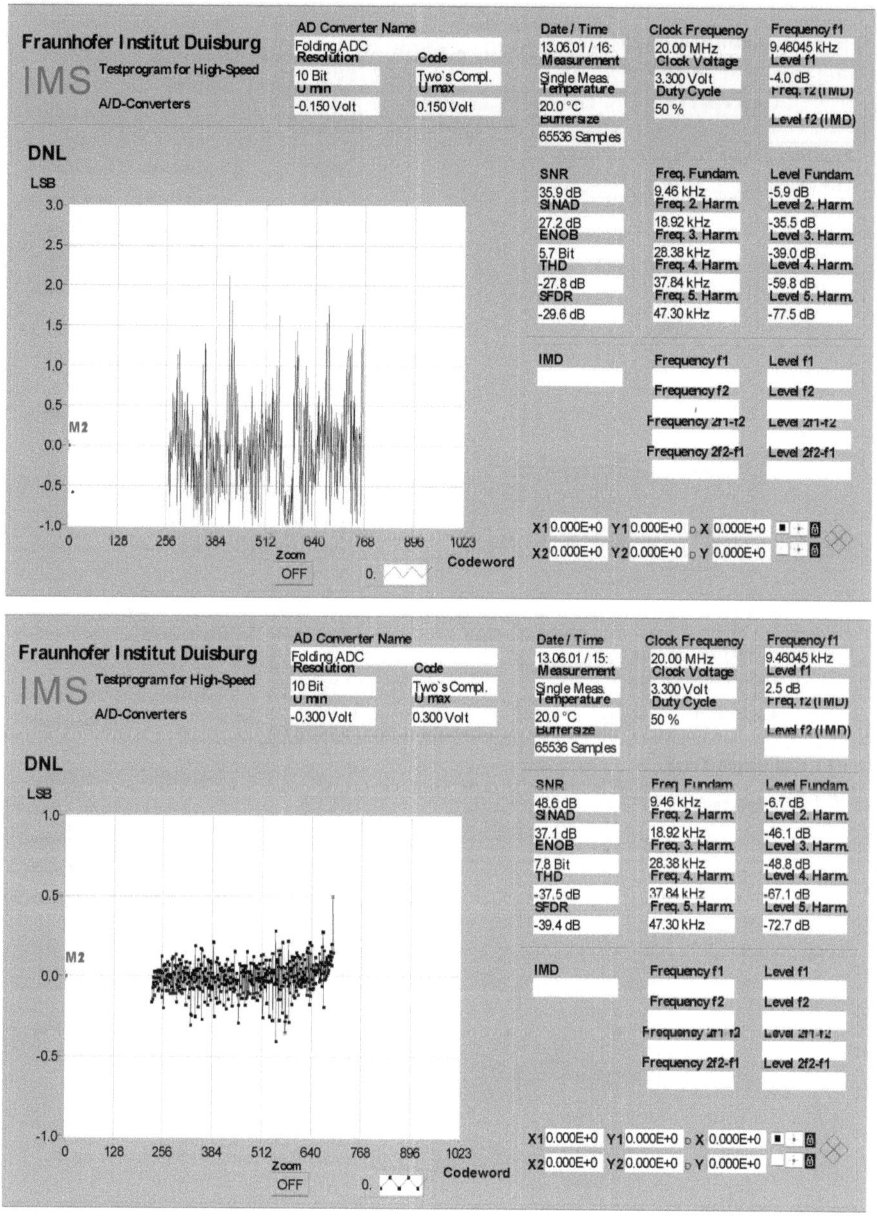

Fig. C.2: DNL analyses of the implemented ADC (above) and the commercial ADC (below)

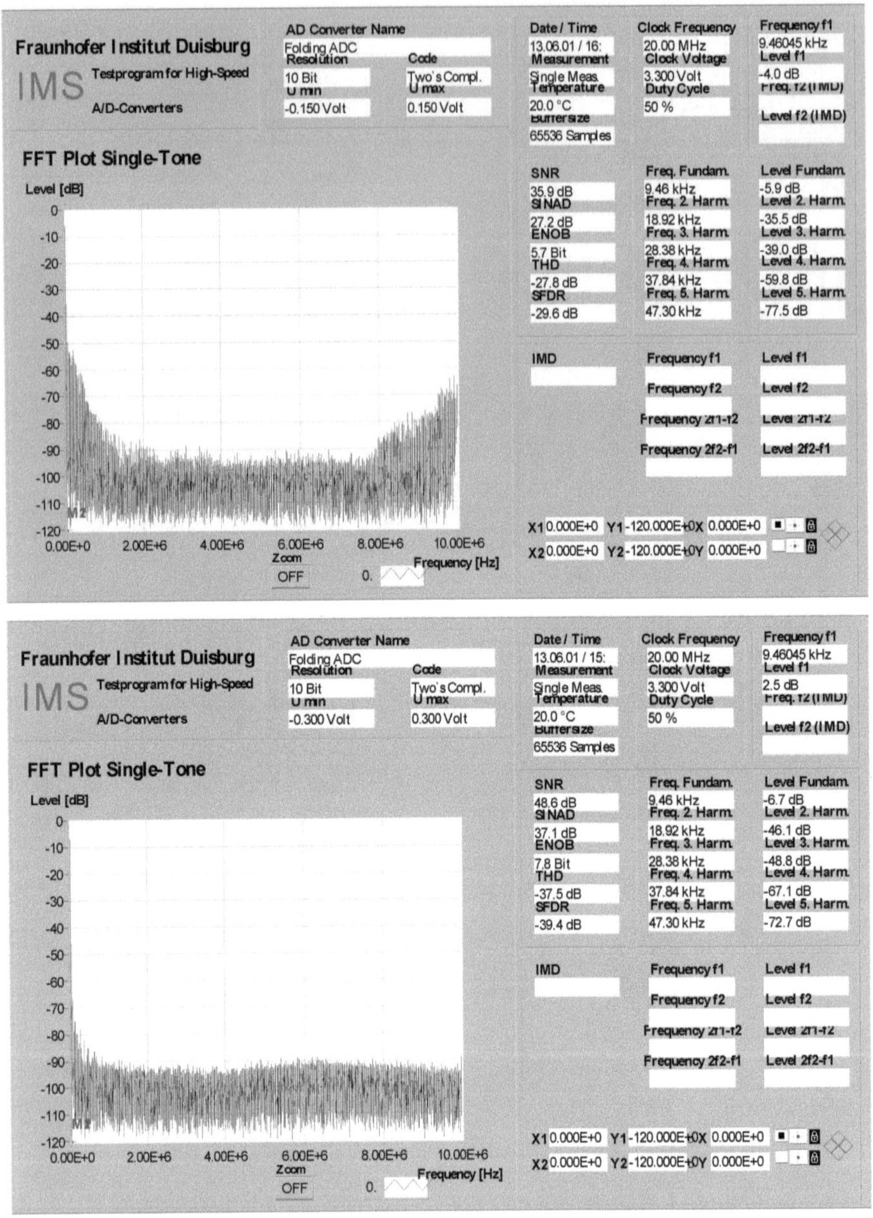

Fig. C.3: FFT analyses at 9.46045 kHz for the implemented ADC (above) and the commercial ADC (below)

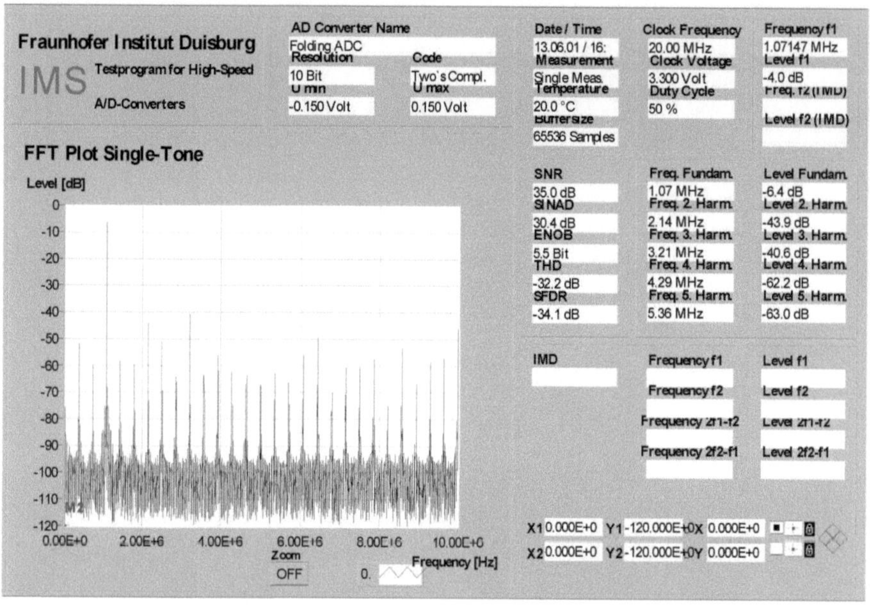

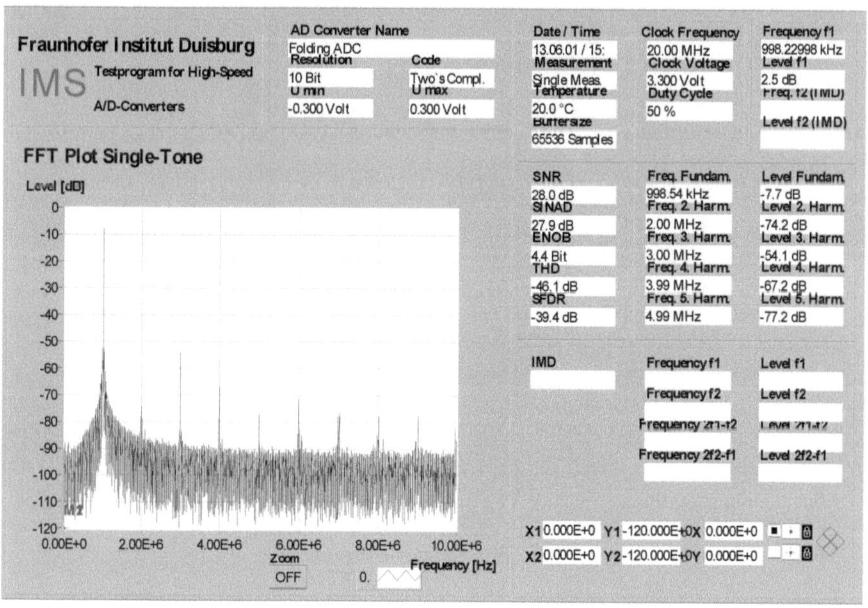

Fig. C.4: FFT analyses at about 1 MHz for the implemented ADC (above) and the commercial ADC (below)

SYMBOLS, CONVENTIONS, NOTATIONS, AND ABBREVIATIONS

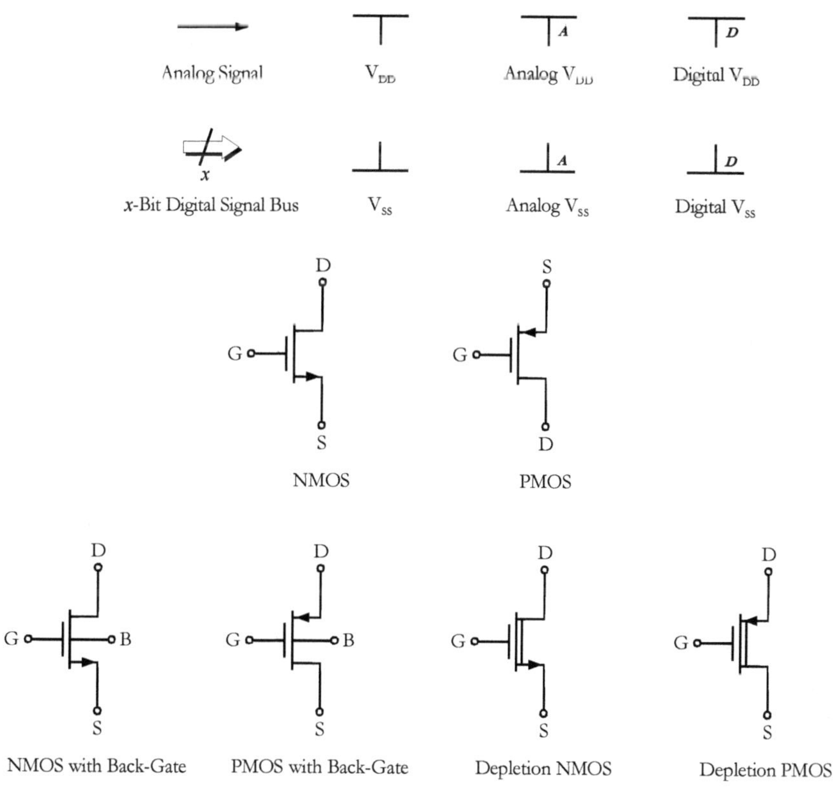

This list gives the full description of the symbols used throughout the text.

A	amplification
A_0	open loop gain, DC gain
A_{CM}	common-mode voltage gain
A_{DM}	differential-mode voltage gain
A_{ITPL}	interstage amplification of the interpolation stage
A_{PP}	peak-to-peak value of the full-scale input signal
A_{Res}	total resolution
A_{RMS}	RMS value of the full-scale input signal
$B_{Eq.}$	equivalent noise bandwidth
C_{GS}	gate-source capacitance
C_H	hold capacitor
C_{In}	input capacitor
C_L	load capacitor
C_M	Miller's capacitor
C_{OX}	gate oxide capacitance per unit area
C_{Para}	parasitic capacitance
D	real transfer function of A/D conversion
$\delta(t)$	Dirac impulse, impulse function
ε	quantization error
ε_{RMS}	root-mean-square value of the quantization error
Φ_M	phase margin
Δf	equivalent noise bandwidth
$f_A(t)$	function of analog signal
f_{0dB}	unity-gain frequency
f_{AD}	A/D conversion rate, A/D conversion frequency
f_{BW}	signal bandwidth
f_C	max. frequency of the input signal
f_{Clk}	clock frequency
$f_D(t)$	function of digital signal
$f_{DA}(t)$	function of the D/A conversion
F_F	folding factor
f_{In}	input frequency
$f_{In}(t)$	input signal
Δf_{In}	bandwidth of the input signal
F_{ITPL}	interpolating factor
f_N	Nyquist frequency
$f_Q(t)$	quantization function
$f_S(t)$	sampling function
FS_j	j^{th} folding signal pair
$FS_{N,j}$	negative j^{th} folding signal
$FS_{P,j}$	positive j^{th} folding signal
Γ	ideal transfer function of A/D conversion; bisection line
g_m	transconductance
$g_{m,\,eff}$	effective transconductance
$H(s)$	transfer function

i current (small signal)

I_{Out} output current

L channel length

LSB lease significant bit

m coarse resolution

MSB most significant bit

MSB_C most significant bit generated by the coarse converter

MSB_{-1} 2nd most significant bit

MSB_{-1CF} 2nd most significant bit generated by the fine thermometer code

MSB_{-2F} 3rd most significant bit generated by the fine converter

n resolution; fine resolution

n_1 folding resolution

n_2 interpolating resolution

$\mathbb{N}$ set of the nature numbers

N total presentable states of an A/D converter

N_{FB} number of folding blocks

N_{FS} number of folding signals

NZ Nyquist zone

P power dissipation

P_{A_ITPL} power dissipation of the amplifier in the interpolation stage

P_{CMP_C} power dissipation of the comparator for the coarse converter

P_{CMP_F} power dissipation of the comparator for the fine converter

P_{FA} power dissipation of the folding amplifier

P_{Ref} power dissipation of the reference resistor

P_{R_ITPL} power dissipation of the resistor in the interpolation stage

$Q(t)$ transfer function for magnitude quantization

Q digital output signal

Θ polar angle

θ_{Diff} phase difference

R_A averaging resistor

R_{DS} drain-source resistor

$\mathrm{rect}(t)$ rectangular pulse

R_{ITPL} interpolating resistor

R_L load resistance

R_{On} on-resistance of CMOS switch

R_{Out} output resistance

S die area; switch

$S_A(t)$ transfer function for analog sampling

S_{A_ITPL} die area of the amplifier in the interpolation stage

S_{CMP} die area of single comparator

S_{CMP_C} die area of the comparator for the coarse converters

S_{CMP_F} die area of the comparator for the fine converters

$S_D(t)$ transfer function for digital sampling

$S_D^{-1}(t)$ inverse transfer function for digital sampling

S_{FA} die area of the folding amplifier

S_{Ref} die area of the reference resistor

S_{R_ITPL} die area of the resistor in the interpolation stage

τ time constant

T period; temperature

T_0 duration of the signal acquisition time

t_{Acq} acquisition time

T_{Clk} clock duration

$TCode$ Thermometer code

t_F fall time

t_H duration for digital "1"

t_L duration for digital "0"

t_{NP} non-overlapped duration

t_P conversion time; propagation delay time

t_R rise time

t_{Set} settling time

V_{Bias} bias voltage

$vbnc$ bias voltage for NMOS transistor

$vbns$ bias voltage for NMOS transistor

$vbpc$ bias voltage for PMOS transistor

$vbps$ bias voltage for PMOS transistor

V_{CFT} voltage of clock feedthrough

V_{DC} DC offset voltage

V_{DD} supply voltage

v_{DS} drain-source voltage (small signal)

V_{DS} drain-source voltage

V_{Error} error source

v_{GS} gate-source voltage (small signal)

V_{GS} gate-source voltage

V_H voltage for digital "1"

v_i difference of the input signals of a differential amplifier

v_{In} input voltage (small signal)

V_{In} input voltage

$V_{In,\,DC}$ input voltage offset

V_{In_max} max. input voltage

V_{In_min} min. input voltage

V_L voltage for digital "0"

V_M virtual ground

$\overline{v_n^2}$ time-dependent error source; random noise source

$\overline{v_{nR}^2}$ time-dependent error source caused (thermal noise) by single resistor

V_{OS} offset voltage

ΔV_{OS} statistical mismatch of the offset voltage

$V_{OS,\,CMP}$ offset voltage of the comparator

$V_{OS,\,Ref}$ offset voltage caused by the reference resistor string

V_{Out} output voltage

$\overline{v_{Out,\,n,\,tot}^2}$ total output noise

V_{Ref} range of the reference voltages

$V_{Ref\_Bus\,j}$ bus of reference voltages

ΔV_{Ref_C} smallest difference among the reference voltages (coarse converter)

ΔV_{Ref_F} smallest difference among the reference voltages (fine converter)

$V_{Ref\_\max}$ max. reference voltage

$V_{Ref\_\min}$ min. reference voltage

V_{SS} ground of the supply voltage

W channel width

Z_{In} input impedance

This list gives the full description of the abbreviations used throughout the text.

A/D	analog-to-digital
AC	alternating current
ADC	analog-to-digital converter
AFE	analog front-end
BER	bit error rate
BPF	bandpass filter
CAD	computer aided design
CFT	clock feedthrough
CMRR	common-mode rejection ratio
CPU	central processing unit
D/A	digital-to-analog
DAC	digital-to-analog converter
dBc	decibels below carrier
dBFS	decibels below full scale
DBS	direct broadcast satellite
DC	direct current
DCS	digitally corrected subranging (ADC)
DFT	discrete Fourier transform
DNL	differential non-linearity
DR	dynamic range
DRAM	dynamic random access memory
DSP	digital signal processor
DS-THA	double-sampled THA
DUT	device under test
DVD	digital versatile disc
ECC	error correction and coding
ENOB	effective number of bits
FA	folding amplifier
FB	folding block
FFT	fast Fourier transform
FS	folding signal
GSPS	giga samples per second
HDTV	high definition TV
IMD	intermodulation distortion
INL	integral non-linearity
LHS	left-hand-side
LPF	lowpass filter
LSB	least significant bit
LSD	least significant digit
MSB	most significant bit
MSD	most significant digit
MSPS	mega samples per second
OSR	oversampling ratio
OTA	operational transconductance amplifier
pp	peak-to-peak
RHS	right-hand-side
RISC	reduced instruction set computers
RMS	root-mean-square
RSS	root-sum-square
S/H	sample-and-hold
S/N	signal-to-noise
SAR	successive approximation register
SC	switched capacitor
SFDR	spurious free dynamic range

SHA sample-and-hold amplifier

SI switched current

SINAD signal-to-noise and distortion ratio

SIP silicon intellectual property

SNDR signal-to-noise and distortion ratio

SNR signal-to-noise ratio

SoC system-on-a-chip

SoS system-on-silicon

SPICE simulation program with IC emphasis

SR slew rate

THA track-and-hold amplifier

THD total harmonic distortion

WER word error rate

INDEX

F

G

H

I

K

L

M

N

O

P

BIBLIOGRAPHY

[1] G. E. Moore, "Cramming more components onto integrated circuits," *Electronics*, vol. 38, no. 8, April 1965.

[2] W. H. Wolf, *Modern VLSI Design: Systems on Silicon*, 2nd ed., New Jersey: Prentice Hall, 1998.

[3] R. Gregorian and G. C. Temes, *Analog MOS Integrated Circuits for Signal Processing*, New York: John Wiley & Sons, 1986.

[4] W. Kester, "High Speed Sampling and High Speed ADCs," in *High Speed Design Techniques*, W. Kester, Ed. Norwood, MA: Analog Devices, Inc., 1996.

[5] H. D. Lüke, *Signalübertragung: Grundlagen der digitalen und analogen Nachrichtenübertragungssysteme*, 5th ed., Berlin, BRD: Springer-Verlag, 1992.

[6] R. Harjani, "Data Converters: Analog-to-Digital Converters," in *The Circuits and Filters Handbook, The Electrical Engineering Handbook Series*, W.-K. Chen, Ed. New York: CRC Press, Inc., 1995.

[7] K.-L. Lin, *Entwicklung von Schaltungstechniken zur Fehlerreduzierung in monolithischen Folding-Analog/Digital-Umsetzern*, Dipl.-Ing. thesis, Duisburg: Gerhard-Mercator-Universität Duisburg, 1998.

[8] I. N. Bronstein and K. A. Semendjajew, *Taschenbuch der Mathematik*, 25th ed., Stuttgart, BRD: B. G. Teubner Verlagsgesellschaft, 1991.

[9] W. Kaplan, *Advanced Mathematics for Engineers*, Reading, MA: Addison-Wesley, 1981.

[10] R. J. v. d. Plassche, *Integrated Analog-to-Digital and Digital-to-Analog Converters*, Boston: Kluwer Academic, 1994.

[11] IEEE, *IEEE Standard 1241 Draft: Standard for Terminology and Test Methods for Analog-to-Digital Converters*, Piscataway, NJ: IEEE, 2000.

[12] D. Seitzer, G. Pretzl, and N. A. Hamdy, *Electronic analog-to-digital converters*, Chichester: John Wiley & Sons Ltd., 1983.

[13] L. Gaddy, "Application Bulletin: Selecting an A/D Converter," Burr-Brown Corp., Tucson, AZ, Application Bulletin AB-098, 1995.

[14] D. F. Hoeschele, *Analog-to-Digital and Digital-to-Analog Conversion Techniques*, 2nd ed., New York: John Wiley & Sons, 1994.

[15] W. Kester, "High Speed A/D Conversion," in *High Speed Design Seminar*, W. Kester, Ed. Norwood, MA: Analog Devices, Inc., 1990.

[16] R. J. v. d. Plassche, "Introduction to High-Speed A/D and D/A Converters," in *Advanced Engineering Course on High-Speed Data Converters*. Lausanne, Switzerland: École Polytechnique Fédérale de Lausanne, 1998.

[17] R. J. v. d. Plassche, "Practical Examples of High-Speed A/D and D/A Conversion in Sub-Micron CMOS," in *Advanced Engineering Course on High-Speed Data Converters*. Lausanne, Switzerland: École Polytechnique Fédérale de Lausanne, 1998.

[18] M. Choi and A. A. Abidi, "A 6 b 1.3 GSample/s A/D converter in 0.35 mu m CMOS," in *IEEE International Solid-State Circuits Conference (ISSCC)*, San Francisco, CA, 2001.

[19] G. Geelen, "A 6 b 1.1 GSample/s CMOS A/D converter," in *IEEE International Solid-State Circuits Conference (ISSCC)*, San Francisco, CA, 2001.

[20] S. Gotoh, T. Takahashi, K. Irie, K. Ohshima, N. Mimura, K. Aida, T. Maeda, T. Yamamoto, K. Sushihara, Y. Okamoto, Y. Tai, T. Nakajima, M. Usui, T. Ochi, K. Komichi, and A. Matsuzawa, "A mixed-signal 0.18 mu m CMOS SOC for DVD systems with 432 MS/s PRML read channel and 16 Mb embedded DRAM," in *IEEE International Solid-State Circuits Conference (ISSCC)*, San Francisco, CA, 2001.

[21] H. v. d. Ploeg, G. Hoogzaad, H. A. H. Termeer, M. Vertregt, and R. L. J. Roovers, "A 2.5 V 12 b 54 MSample/s 0.25 mu m CMOS ADC in 1 mm/sup 2," in *IEEE International Solid-State Circuits Conference (ISSCC)*, San Francisco, CA, 2001.

[22] R. C. Taft and M. R. Tursi, "A 100-MS/s 8-b CMOS subranging ADC with sustained parametric performance from 3.8 V down to 2.2 V," *IEEE Journal of Solid-State Circuits,* vol. 36, no. 3, pp. 331-338, March 2001.

[23] L. Singer, "High-Speed Pipelined ADC Architectures," in *Advanced Engineering Course on High-Speed Data Converters*. Lausanne, Switzerland, 1998.

[24] H.-S. Chen, B.-S. Song, and K. Bacrania, "A 14-b 20-Msamples/s CMOS pipelined ADC," *IEEE Journal of Solid-State Circuits,* vol. 36, no. 6, pp. 997-1001, June 2001.

[25] A. Gerosa, R. Bernardini, and S. Pietri, "A fully integrated 8-bit, 20 MHz, truly random numbers generator, based on a chaotic system," in *Southwest Symposium on Mixed-Signal Design*, Austin, TX, 2001.

[26] H. Liu and M. Hassoun, "High speed re-configurable pipeline ADC cell design," in *Southwest Symposium on Mixed-Signal Design,* Austin, TX, 2001.

[27] D. Miyazaki and S. Kawahito, *A high-speed low-power area-efficient pipeline A/D converter and its design method,* Report, Hamamatsu, Japan: Shizuoka University, 2001.

[28] B. Nejaati, A. Khakifirooz, S. J. Ashtiani, and O. Shoaei, "Pipeline analog-to-digital converters with radix <2," in *The 12th International Conference on Microelectronics (ICM),* Tehran, Iran, 2000.

[29] Y.-I. Park, S. Karthikeyan, F. Tsay, and E. Bartolome, "A 10 b 100 MSample/s CMOS pipelined ADC with 1.8 V power supply," in *IEEE International Solid-State Circuits Conference (ISSCC),* San Francisco, CA, 2001.

[30] P. P. Siniscalchi, J. K. Pitz, R. K. Hester, S. M. DeSoto, W. Minsheng, S. Sridharan, R. L. Halbach, D. Richardson, W. Bright, M. M. Sarraj, J. R. Hellums, C. L. Betty, and G. Westphal, "A CMOS ADSL codec for

central office applications," *IEEE Journal of Solid-State Circuits*, vol. 36, no. 3, pp. 356-365, March 2001.

[31] M. Waltari and K. A. I. Halonen, "1-V 9-bit pipelined switched-opamp ADC," *IEEE Journal of Solid-State Circuits*, vol. 36, no. 1, pp. 129-134, January 2001.

[32] H. Pan, M. Segami, M. Choi, L. Cao, and A. A. Abidi, "A 3.3-V 12-b 50-MS/s A/D converter in 0.6- mu m CMOS with over 80-dB SFDR," *IEEE Journal of Solid-State Circuits*, vol. 35, no. 12, pp. 1769-1780, December 2000.

[33] C. Shi, Y. Wu, C.-H. Lin, S. Li, and M. Ismail, "Design and power optimization of high-speed pipeline ADC for wideband CDMA applications," in *The 17th NORCHIP Conference*, Oslo, Norway, 1999.

[34] H. Bilhan and M. W. Gosney, "A 13 bit 20 MS/s current mode pipelined analog to digital converter," in *The 42nd Midwest Symposium on Circuits and Systems*, Las Cruces, NM, 1999.

[35] S. Imai and Y. Sugimoto, "A design of a bit-block circuit applicable to a 1 V operational video-rate current-mode CMOS A/D converter," *Transactions of the Institute of Electrical Engineers of Japan, Part C,* vol. 120-C, no. 10, pp. 1325-1332, October 2000.

[36] D. G. Nairn, "A 10-bit, 3 V, 100 MS/s pipelined ADC," in *IEEE Custom Integrated Circuits Conference (CICC)*, Orlando, FL, 2000.

[37] I. Mehr and L. Singer, "A 55-mW, 10-bit, 40-Msample/s Nyquist-rate CMOS ADC," *IEEE Journal of Solid-State Circuits*, vol. 35, no. 3, pp. 318-325, March 2000.

[38] I. Galton, "Digital cancellation of D/A converter noise in pipelined A/D converters," *IEEE Transactions on Circuits & Systems II-Analog & Digital Signal Processing*, vol. 47, no. 3, pp. 185-196, March 2000.

[39] D. Kelly, W. Yang, I. Mehr, M. Sayuk, and L. Singer, "A 3 V 340 mW 14 b 75 MSPS CMOS ADC with 85 dB SFDR at Nyquist," in *IEEE International Solid-State Circuits Conference (ISSCC)*, San Francisco, CA, 2001.

[40] B. Razavi, "Folding and Interpolating ADCs," in *Advanced Engineering Course on High-Speed Data Converters*. Lausanne, Switzerland: École Polytechnique Fédérale de Lausanne, 1998.

[41] W. T. Colleran and A. A. Abidi, "A 10-b, 75-MHz two-stage pipelined bipolar A/D converter," *IEEE Journal of Solid-State Circuits*, vol. 28, no. 12, pp. 1187-1199, December 1993.

[42] R. E. J. v. d. Grift and R. J. v. d. Plassche, "A monolithic 8-bit video A/D converter," in *9th European Solid-State Circuits Conference (ESSCIRC)*, Lausanne, Switzerland, 1983.

[43] R. E. J. v. d. Grift and R. J. v. d. Plassche, "A monolithic 8-bit video A/D converter," *IEEE Journal of Solid-State Circuits*, vol. SC-19, no. 3, pp. 274-278, June 1984.

[44] R. E. J. v. d. Grift, I. W. J. M. Rutten, and M. v. d. Veen, "An 8-bit video ADC incorporating folding and interpolation techniques," *IEEE Journal of Solid-State Circuits*, vol. SC-22, no. 6, pp. 944-953, December 1987.

[45] R. J. v. d. Plassche and P. Baltus, "An 8-bit 100-MHz full-Nyquist analog-to-digital converter," *IEEE Journal of Solid-State Circuits*, vol. 23, no. 6, pp. 1334-1344, December 1988.

[46] J. v. Valburg and R. J. v. d. Plassche, "An 8-b 650-MHz folding ADC," *IEEE Journal of Solid-State Circuits*, vol. 27, no. 12, pp. 1662-1666, December 1992.

[47] M. P. Flynn and D. J. Allstot, "CMOS folding ADCs with current-mode interpolation," in *IEEE International Solid-State Circuits Conference (ISSCC)*, San Francisco, CA, 1995.

[48] K. L. Lin, T. van den Boom, N. Stevanović, J. Driesen, D. Hammerschmidt, and B. Hosticka, "A basic design guide for CMOS folding and interpolating A/D converters-overview and case study," in *The 6th IEEE International Conference on Electronics, Circuits and Systems (ICECS)*, Pafos, Cyprus, 1999.

[49] K.-M. Kim and K.-S. Yoon, "An 8-bit 42 Msamples/s current-mode folding and interpolation CMOS analog-to-digital converter with three-level folding amplifiers," in *The 39th Midwest Symposium on Circuits and Systems*, Ames, IA, 1996.

[50] K.-M. Kim and K.-S. Yoon, "An 8-bit CMOS current-mode folding and interpolation A/D converter with three-level folding amplifiers," in *IEEE Asia Pacific Conference on Circuits and Systems (APCCAS)*, Seoul, South Korea, 1996.

[51] M. P. Flynn and B. Sheahan, "A 400-Msample/s, 6-b CMOS folding and interpolating ADC," *IEEE Journal of Solid-State Circuits*, vol. 33, no. 12, pp. 1932-1938, December 1998.

[52] M. Flynn and B. Sheahan, "A 400 M sample/s 6b CMOS folding and interpolating ADC," in *IEEE International Solid-State Circuits Conference (ISSCC)*, San Francisco, CA, 1998.

[53] X. Jiang, Y. Wang, and A. N. Willson, Jr., "A 200 MHz 6-bit folding and interpolating ADC in 0.5-mu m CMOS," in *IEEE International Symposium on Circuits and Systems (ISCAS)*, Monterey, CA, 1998.

[54] K.-M. Kim and K.-S. Yoon, "An 8 bit current-mode CMOS A/D converter with three level folding amplifiers," *IEICE Transactions on Fundamentals of Electronics Communications & Computer Sciences*, vol. E81-A, no. 2, pp. 252-255, February 1998.

[55] H.-H. Kim and K.-S. Yoon, "A 12 bit current-mode folding/interpolation CMOS A/D converter with multipliers," in *IEEE Region 10 Conference (TENCON 99) 'Multimedia Technology for Asia-Pacific Information Infrastructure'*, Cheju Island, South Korea, 1999.

[56] K.-M. Kim and K. S. Yoon, "An 8 bit current-mode CMOS A/D converter with three level folding amplifiers," *Analog Integrated Circuits & Signal Processing*, vol. 20, no. 2, pp. 139-143, August 1999.

[57] M.-J. Choe, B.-S. Song, and K. Bacrania, "An 8 b 100 MSample/s CMOS pipelined folding ADC," in *Symposium on VLSI Circuits*, Kyoto, Japan, 1999.

[58] H.-H. Kim and K.-S. Yoon, "A 12 bit current-mode folding/interpolation CMOS A/D converter with 2 step architecture," in *The 1st IEEE Asia Pacific Conference on ASICs (AP-ASIC)*, Seoul, South Korea, 1999.

[59] A. Pierazzi and A. Boni, "Design issues for high frequency, 0.35 mu m, 3.3 V CMOS folding A/D converter," in *The 3rd International Con-*

ference on Advanced A/D and D/A Conversion Techniques and their Applications, Glasgow, UK, 1999.

[60] J.-J. Moon and M.-K. Song, "Design of a low power 3 V 6-bit 100 MSPS CMOS ADC for DBS receiver," *Journal of the Institute of Electronics Engineers of Korea C*, vol. 36-C, no. 12, pp. 20-26, December 1999.

[61] K. Yoon, J. Lee, D.-K. Jeong, and W. Kim, "An 8-bit 125 MS/s CMOS folding ADC for Gigabit Ethernet LSI," in *Symposium on VLSI Circuits*, Honolulu, HI, 2000.

[62] T.-H. Kim, J.-J. Sung, S.-H. Kim, W. Joo, S.-B. You, and S. Kim, "A 10-bit, 40 Msamples/s cascading folding and interpolating A/D converter with wide range error correction," in *The 2nd IEEE Asia Pacific Conference on ASICs (AP-ASIC)*, Cheju, South Korea, 2000.

[63] J.-W. Chung and K.-S. Yoon, "Design of 3.3 V 10 bit current-mode folding/interpolating CMOS A/D converter with an arithmetic functionality," in *The 2nd IEEE Asia Pacific Conference on ASICs (AP-ASIC)*, Cheju, South Korea, 2000.

[64] A. R. Nabavi and K. Dabbagh, "A 10-bit, 20 Ms/s, 22 mW folding and interpolating CMOS ADC," in *The 12th International Conference on Microelectronics (ICM)*, Tehran, Iran, 2000.

[65] M.-J. Choe, B.-S. Song, and K. Bacrania, "A 13 b 40 MSample/s CMOS pipelined folding ADC with background offset trimming," in *IEEE International Solid-State Circuits Conference (ISSCC)*, San Francisco, CA, 2000.

[66] M.-J. Choe, B.-S. Song, and K. Bacrania, "A 13-b 40-MSamples/s CMOS pipelined folding ADC with background offset trimming," *IEEE Journal of Solid-State Circuits*, vol. 35, no. 12, pp. 1781-1790, December 2000.

[67] H.-H. Kim and K.-S. Yoon, "A current-mode folding/interpolating CMOS A/D converter with multiplied folding amplifiers," *IEICE Transactions on Fundamentals of Electronics Communications & Computer Sciences*, vol. E84-A, no. 2, pp. 563-567, February 2001.

[68] M.-J. Choe, B.-S. Song, and K. Bacrania, "An 8-b 100-MSample/s CMOS pipelined folding ADC," *IEEE Journal of Solid-State Circuits,* vol. 36, no. 2, pp. 184-194, February 2001.

[69] M.-H. Liu and S.-I. Liu, "An 8-bit 10 MS/s folding and interpolating ADC using the continuous-time auto-zero technique," *IEEE Journal of Solid-State Circuits,* vol. 36, no. 1, pp. 122-128, January 2001.

[70] M. P. Flynn and D. J. Allstot, "CMOS folding A/D converters with current-mode interpolation," *IEEE Journal of Solid-State Circuits,* vol. 31, no. 9, pp. 1248-1257, September 1996.

[71] K. Bult and A. Buchwald, "An embedded 240-mW 10-b 50-MS/s CMOS ADC in 1-mm^2," *IEEE Journal of Solid-State Circuits,* vol. 32, no. 12, pp. 1887-1895, December 1997.

[72] K. Bult, A. Buchwald, and J. Laskowski, "A 170mW 10b 50MSamples/s CMOS ADC in 1-mm^2," in *IEEE International Solid-State Circuits Conference (ISSCC),* San Francisco, CA, 1997.

[73] K. Bult, "High-Speed ADCs," in *Advanced Engineering Course on High-Speed Data Converters.* Lausanne, Switzerland: École Polytechnique Fédérale de Lausanne, 1998.

[74] B. Nauta and A. G. W. Venes, "A 70-MS/s 110-mW 8-b CMOS folding and interpolating A/D converter," *IEEE Journal of Solid-State Circuits,* vol. 30, no. 12, pp. 1302-1308, December 1995.

[75] B. Nauta and A. G. W. Venes, "A 70 MSample/s 110 mW 8 b CMOS folding interpolating A/D Converter," in *IEEE International Solid-State Circuits Conference (ISSCC),* San Francisco, CA, 1995.

[76] B. Brandt, "Subranging CMOS ADCs," in *Advanced Engineering Course on High-Speed Data Converters.* Lausanne, Switzerland, 1998.

[77] B. Razavi, *Principles of Data Conversion System Design,* New York: IEEE Press, 1995.

[78] B. Razavi, "High-Speed Sample-and-Hold Circuits," in *Advanced Engineering Course on High-Speed Data Converters.* Lausanne, Switzerland: École Polytechnique Fédérale de Lausanne, 1998.

[79] B. J. Hosticka, *personal communication,* September, 2001.

[80] M. Waltari and K. Halonen, "A 10-bit 220-MSample/s CMOS sample-and-hold circuit," in *IEEE International Symposium on Circuits and Systems (ISCAS)*, Monterey, CA, 1998.

[81] M. Waltari and K. Halonen, "A 220-Msample/s CMOS sample-and-hold circuit using double-sampling," *Analog Integrated Circuits & Signal Processing*, vol. 18, no. 1, pp. 21-31, January 1999.

[82] Y. M. Lin, B. Kim, and P. R. Gray, "A 13-b 2.5-MHz self-calibrated pipelined A/D converter in 3- mu m CMOS," *IEEE Journal of Solid-State Circuits*, vol. 26, no. 4, pp. 628-636, April 1991.

[83] S. Sheng and R. Brodersen, *Low-Power CMOS Wireless Communications: A Wideband CDMA System Design*, Bosten: Kluwer Academic Publishers, 1998.

[84] K. Bult and G. J. G. M. Geelen, "The CMOS gain-boosting technique," *Analog Integrated Circuits and Signal Processing*, vol. 1, no. 2, pp. 119-135, October 1991.

[85] B. J. Hosticka, "Improvement of the gain of MOS amplifiers," *IEEE Journal of Solid-State Circuits*, vol. SC-14, no. 6, pp. 1111-1114, December 1979.

[86] P. R. Gray, P. J. Hurst, S. H. Lewis, and R. G. Meyer, *Analysis and Design of Analog Integrated Circuits*, 4th ed., New York: John Wiley & Sons, Inc., 2001.

[87] F. J. Tegude, "Skript zur Vorlesung: Technische Elektronik 1 (Feldeffekt-Elektronik)," vol. 1. Duisburg, BRD: Fachbereich Elektrotechnik, Gerhard-Mercator-Universität Duisburg, 1993.

[88] U. Tietze and C. Schenk, *Halbleiter-Schaltungstechnik*, 11th ed., Berlin: Springer-Verlag, 1999.

[89] K. R. Laker and W. M. C. Sansen, *Design of Analog Integrated Circuits and Systems*, New York: McGraw-Hill, 1994.

[90] B. J. Hosticka, *personal communication*, August, 2001.

[91] K. Kattman and J. Barrow, "A technique for reducing differential non-linearity errors in flash A/D converters," in *IEEE International Solid-State Circuits Conference (ISSCC)*, San Francisco, CA, 1991.

[92] S. Wenzek, *Entwicklung eines monolithischen Folding-Analog/Digital-Konverters in CMOS-Technologie,* Dipl.-Ing. thesis, Duisburg: Gerhard-Mercator-Universität Duisburg, 1998.

[93] G. M. Yin, F. O. Eynde, and W. Sansen, "A high-speed CMOS comparator with 8-b resolution," *IEEE Journal of Solid-State Circuits,* vol. 27, no. 2, pp. 208-211, February 1992.

[94] W. C. Slemmer, "High-speed low-power strobed comparator," *IEEE Journal of Solid-State Circuits,* vol. SC-5, no. 5, pp. 215-220, October 1970.

[95] H. L. Fiedler, B. Hoefflinger, W. Demmer, and P. Draheim, "A 5-bit building block for 20 MHz A/D converters," *IEEE Journal of Solid-State Circuits,* vol. SC-16, no. 3, pp. 151-155, June 1981.

[96] J.-H. Shieh, M. Patil, and B. J. Sheu, "Measurement and analysis of charge injection in MOS analog switches," *IEEE Journal of Solid-State Circuits,* vol. SC-22, no. 2, pp. 277-281, April 1987.

[97] G. Wegmann, E. A. Vittoz, and F. Rahali, "Charge injection in analog MOS switches," *IEEE Journal of Solid-State Circuits,* vol. SC-22, no. 6, pp. 1091-1097, December 1987.

[98] C. Eichenberger and W. Guggenbuhl, "On charge injection in analog MOS switches and dummy switch compensation techniques," *IEEE Transactions on Circuits and Systems,* vol. 37, no. 2, pp. 256-264, February 1990.

[99] B. Jonsson and N. Tan, "Clock-feedthrough compensated first-generation SI circuits and systems," *Analog Integrated Circuits & Signal Processing,* vol. 12, no. 3, pp. 201-210, April 1997.

[100] M. F. Li, S. Y. Yep, and Y. C. Lim, "A novel integrated CMOS switch circuit for high precision sample-and-hold technique," *Analog Integrated Circuits & Signal Processing,* vol. 12, no. 3, pp. 211-215, April 1997.

[101] R. H. Walden, "Analog-to-digital converter survey and analysis," *IEEE Journal on Selected Areas in Communications,* vol. 17, no. 4, pp. 539-550, April 1999.

[102] W. Geisselhardt and A. Hunger, "Vorlesungsskripte: Datenverarbeitung," vol. 3. Duisburg, BRD: Fachbereich Elektrotechnik, Gerhard-Mercator-Universität Duisburg, 1996.

[103] W. Kester, J. Bryant, W. Jung, A. Garcia, J. McDonald, and J. Buxton, "High Speed Hardware Design Techniques," in *High Speed Design Techniques*, W. Kester, Ed. Norwood, MA: Analog Devices, Inc., 1996.

[104] T. v. d. Boom, *personal communication*, August, 2000.

[105] G. R. Boyle, B. M. Cohn, D. O. Pederson, and J. E. Solomon, "Macromodeling of Integrated Circuit Operational Amplifiers," *IEEE Journal of Solid-State Circuits*, vol. SC-9, no. 6, pp. 353-364, June 1974.

[106] R. Wittmann, W. Schardein, B. J. Hosticka, G. Burbach, and J. Arndt, "Trimless high precision ratioed resistors in D/A and A/D converters," *IEEE Journal of Solid-State Circuits*, vol. 30, no. 8, pp. 935-939, August 1995.

[107] J. Choma, Jr., "Monolithic Device Models: MOSFET Technology Devices," in *The Circuits and Filters Handbook, The Electrical Engineering Handbook Series*, W.-K. Chen, Ed. New York: CRC Press, Inc., 1995.

[108] D. K. Su, M. J. Loinaz, S. Masui, and B. A. Wooley, "Experimental results and modeling techniques for substrate noise in mixed-signal integrated circuits," *IEEE Journal of Solid-State Circuits*, vol. 28, no. 4, April 1993.

[109] T. J. Schmerbeck, "Practical Aspects of Analogue and Mixed-Mode ICs," Lausanne, Switzerland: École Polytechnique Fédérale de Lausanne, 2002.

[110] Amkor Technology, "fcCSP flipChip CSP package data sheet," http://www.amkor.com.

[111] Amkor Technology, "fcMLF Micro LeadFrame™ data sheet," http://www.amkor.com.

[112] B. R. Stanisic, N. K. Verghese, R. A. Rutenbar, L. R. Carley, D. J. Allstot, "Adressing Substrate Coupling in Mixed-Mode IC's: Simulation and Power Distribution Synthesis," *IEEE Journal of Solid-State Circuits*, vol. 29, no. 3, March 1994.

[113] M. v. Heijningen, J. Compiet, P. Wambacq, S. Donnay, M. G. E. Engels, I. Bolsens, "Analysis and Experimental Verification of Digital Substrate Noise Generation for Epi-Type Substrates," *IEEE Journal of Solid-State Circuits*, vol. 35, no. 7, July 2000.

[114] N. K. Verghese, T. J. Schmerbeck, D. J. Allstot, *Simulation Techniques and Solutions for Mixed-Signal Coupling in Integrated Circuits*, Bosten: Kluwer Academic Publishers, 1995.

[115] V. S. Gierenz, R. Schwann, and T. G. Noll, "A low power digital beamformer for handheld ultrasound systems," in *27th European Solid-State Circuits Conference (ESSCIRC)*, Villach, Austria, 2001.

MIX
Papier aus verantwortungsvollen Quellen
Paper from responsible sources
FSC® C105338

If you have any concerns about our products,
you can contact us on
ProductSafety@springernature.com

In case Publisher is established outside the EU,
the EU authorized representative is:
Springer Nature Customer Service Center GmbH
Europaplatz 3, 69115 Heidelberg, Germany

Printed by Libri Plureos GmbH
in Hamburg, Germany